DIGITAL
HARMONY

Also by Arno Penzias

Ideas and Information

DIGITAL HARMONY

BUSINESS, TECHNOLOGY
&
LIFE
AFTER
PAPERWORK

ORIGINALLY PUBLISHED UNDER THE TITLE *HARMONY*

ARNO PENZIAS

HarperBusiness
A Division of HarperCollinsPublishers

First paperback edition published 1996.

Designed by Nancy Singer

The Library of Congress has catalogued the hardcover edition as follows:

Penzias, Arno A.
    Harmony : business, technology & life after paperwork / Arno Penzias. — 1st ed.
      p. cm.
    Includes index.
    ISBN 0-88730-724-8
    1. Automation—Management. 2. Business enterprises—Technological
innovations. 3. Paperwork (Office practice)—Technological innovations. I. Title.
HD45.2.P46 1995
651.8—dc20   94-47435

ISBN 0-88730-785-X (pbk.)

96 97 98 99 00 ❖/RRD 10 9 8 7 6 5 4 3 2 1

To Karl and Justine Penzias,
the hardworking immigrants who made it all possible
With love and gratitude

CONTENTS

ACKNOWLEDGMENTS

This book's organizing themes owe much to Bill Leigh's sound advice and insightful criticism. Starting with a perplexing collection of observations about the interplay between technology and business, Bill read and helped redirect each of my early attempts to capture the essence of this story.

Warm thanks are also due to Adrian Zackheim for his belief in a fledgling concept, and the energetic enthusiasm with which he guided it to fruition. Thanks also to Sally Arteseros, who helped give the manuscript a "haircut," as she called it, prior to its final submission for publication.

As before, Julia Heiney managed to fit all the extra work that a book entails into her more than full-time job of keeping everything that moves through my office (including me) running smoothly.

Finally, I'm grateful to the friends, colleagues, and business acquaintances who gave generously of their time and insight to help shape the ideas and supporting material presented herein. While I can't possibly list them all, I would like to make special mention of Greg Blonder, Martin Fransman, Sandy Fraser, Peter Keene, Sherry Levit, Arun Netravali, Tony Ottinger, Larry Rabiner, and Rick Wright. Many thanks to all.

INTRODUCTION

Fast and far-reaching change unsettles every corner of our global economy. So much so that traditional signs of economic recovery give little comfort to its supposed beneficiaries. What, after all, can entrepreneurs, managers, and employees really count on in this era of jobless prosperity?

For decades, office managers around the world have cried for greater efficiency—only to find that an alarming number of tasks now require no labor at all. In their place, information users and suppliers merely communicate with one another and bypass all the steps along the way. While millions of us still shuffle paper for a living, no one can ignore the growing signs of change. Just look at our latest tools. As more information flows through fax machines, bar-code readers, ATMs, and electronic mail, it passes through fewer human hands. Small wonder then that office workers everywhere wonder about their futures.

For most of us, the Information Revolution's negative aspects seem far more evident than its benefits. Business and government leaders hail the advent of Information Superhighways but stumble on specifics. What new products and services will attract customers, reward investors, and provide our workforce with meaningful sources of new employment? Clearly, few of us will wish to wait passively until one or another scenario plays itself out.

Investors, jobholders, and job seekers will surely want to manage their money and careers on the basis of insight rather than chance. Consumers will benefit from understanding what technology brings to the marketplace and how it may impact upon their lives. Concerned citizens will need a clear view of what lies ahead for themselves and their children. Technology and its interplay with business and economics seems too important to remain in the domain of technologists alone—especially when revolution is in the air.

What can we say with certainty about the connection between computerization and job shrinkage? While white-collar layoffs abound, no one talks about office automation anymore. Even the most outspoken Artificial Intelligence enthusiast wouldn't claim that a computer could perform the eclectic array of tasks that a purchasing agent, a financial analyst, or a regional sales manager performs in a typical workday. And yet formerly busy people in many such walks of life find themselves unemployed—while the companies for which they once worked continue on without their services. While a computer can rarely replace a human office worker in a given job, it seems, streamlined information flows can make an office full of jobs unnecessary.

What is happening? As I see it, traditional computers have, as yet, done little to alter the pace of office work. Just look at the white-collar job growth that has accompanied the widespread introduction of computers since the 1970s. But far-reaching changes lie below the surface. In particular, the overwhelming majority of our significant record keeping—in banking, billing, travel reservations, and cataloging, for example—has now been computerized. And since most of the transactions shaping our lives now pass through at least one computer as they move toward completion, a large fraction of the world's tasks boil down to issuing commands to the right computer.

In today's world, getting information to and from such computers generally requires multiple layers of human intermediaries. These intermediaries—clerks, middle managers, stenographers, directory-assistance operators, travel agents, and the like—constitute a large portion of the group we now call "office workers." Few of them started out in this computer-interface role. Instead, the computers grew up around them.

Consider a bank on the day after it switched its account records to a computer. From that time onward, a customer's deposit had to be recorded by altering a number stored in that computer's memory. To expedite the process, the bank hired a data-entry clerk to translate deposit slips into computer-readable symbols. But that didn't last long. After a while, tellers punched the deposited amounts into terminals themselves, thereby assuming the data-entry clerk's function. And then ATMs came along—and circumvented every job between a customer and the bank's records.

This circumvention mechanism lies at the heart of a host of seemingly unrelated phenomena—such as organizational flattening and the negative cost of quality, for example. In the latter instance, it costs *less money* to make defect-free VCRs than it does to produce less perfect ones. And as we'll see later, this aspect of the computing/communication revolution allowed Japan's information-rich factories to vault past their entrenched American competitors.

While U.S. producers have regained at least a portion of their lost ground, their revamped enterprises invariably require fewer workers. And therein lies the challenge. With fewer people needed to produce goods and services, keeping our workforce fully employed requires a steady stream of new offerings. Where will these new opportunities come from? With enormous revenue streams at stake, most established businesses struggle to maintain their share of existing markets—markets already crowded with similar offerings.

As I see it, our most promising opportunities lie at the peripheries of today's products and services—and in the gaps between them. In 1985, for instance, a young couple started an electronics business in their garage. Ten years later, Wall Street valued their Cisco Systems at over $5 billion. Why? While dozens of established computer companies vied with one another to produce equipment that would attach to networks, the Cisco pioneers crafted a device to tie networks together. Anxious to sell their existing products, Cisco's far larger competitors turned a blind eye to a hugely profitable shortcoming in their technology.

Looking to the future, we can expect tomorrow's success stories to grow from the seeds of current shortcomings. And shortcomings abound. While most of today's "high-quality" products work exactly as their designers intended, many of their features baffle the average user. Annoying gaps persist between systems as well, obliging humans to act as messengers between them. Just look at a modern hospital. "High-tech" machinery is everywhere, yet physicians and nurses must rummage through scruffy stacks of paper at every turn. Finally, present-day technology frequently puts economic and environmental considerations in opposition to one another. Serious shortcomings mark this arena as well.

Herein lies an answer to the paradox of our jobless prosperity. Since true productivity implies the creation of value, we must remember that products and services exist to improve the quality of human life rather than as ends in themselves. Stuck in the old paradigm, too many products—and too many producers—compete in glutted markets. Instead, we can expand our horizons to recognize unmet needs for products and services.

I believe that we are already beginning to move in this direction. I see this new period as a time of harmony, in which tomorrow's businesses will produce offerings that work *with one another*, that suit our global environment, and that adapt themselves to humans (instead of the other way around). I believe that in this new Era of Harmony we can provide useful employment for all the skills and talents a healthy society can muster.

I hope this book will aid in that endeavor—by describing the problems we face, the technologies available to us, and the opportunities they make possible.

CHAPTER 1

REVOLUTION

In the short span of three human lifetimes, advances in technology have driven three revolutions—in modern agriculture, in manufacturing, and in information. And like some staggering prizefighter, our economy reels from the Information Revolution's one-two punch. First the factory and now the office. With the onset of computer-controlled manufacturing in the 1970s, industry after industry faced massive layoffs, and jobs moved to better-organized factories on the Pacific Rim.

Offices seemed immune at first, but that situation changed markedly in the early 1990s. This time, however, the jobs didn't shift from one region to another; instead, they moved from people to the computers that served them. New means of direct access to information has made cumbersome paperwork unnecessary, so companies find themselves able to shed employees by the thousands.

Faced with present-day turmoil, many yearn for a return to simpler times. But they yearn in vain. Revolution-driven change runs too deep. Can anyone imagine a return to the scruffy imperfections of a 1960s TV set or automobile? Of course not. Today's consumers take *quality* for granted. And in the same way new marketplace realities now impose an equally dramatic change on the output of office work—a value that I call *harmony*.

We will be exploring the aspects of this Era of Harmony in the following chapters. I see it as a time in which business and technology will work together in greater unity, a "postpaperwork" period in which companies will look outward to their customers instead of inward to their own bureaucracies. As a result, a period of new and exciting opportunities lies ahead of us.

After decades of stagnation, middle managers do seem more productive today—at least those who still have jobs. Having barely adjusted to the challenge of high-quality manufacture from the Pacific Rim, today's workforce now faces further upheaval from the explosive growth of electronic commerce. In a networked world, business must move at the speed of light or risk getting left behind.

Engaged in head-to-head competition to maintain their share of largely stagnant markets, change-resistant contenders seek to boost profits by cutting costs. And that generally boils down to getting more output out of fewer workers.

Yet others see technology as a stepping-stone into new markets rather than just a way to maintain themselves in their present businesses. While only a handful of these eager entrepreneurs reach billion-dollar status, millionaires abound. Both a tool for disemployment and an engine for new economic opportunities, technology has changed the workplace dramatically. And if the following present-day example reflects what lies in store for our economy, we ain't seen nothin' yet.

Information Access

Much change lies ahead for offices still mired in paperwork. When a friend of mine opened a fuel oil business in Harrisburg, Pennsylvania, some thirty years ago, loading a truck at the wholesale supplier involved a wad of multilayer paper forms.

In those days, my friend's drivers would begin each day's route with a fill-up at the local Esso distribution center. Stopping first at the center's office, each driver would get a partially completed multicarbon form from the clerk, then proceed to the pump (where the operator would record on the form the amount of oil dispensed at the end of each fill-up) and then back to the office again. Numbers entered, receipts signed, carbons torn off, duplicate records made and checked to guard against fraud and error—I'm sure you get the idea.

Today, my friend's drivers still deliver the same brand of fuel to many of the same homes as they did in the sixties, and still pick up the oil at the same distribution center. But the center has changed. That Exxon facility, as it's now called, no longer needs an office. In fact, the entire property requires just one employee—a safety inspector who guards against misuse.

Each driver merely pulls up to a pump, hooks up the hose, and slips his company's card into a slot to start the flow. When done, the driver retrieves the card and waits a few seconds while a nearby printer creates a bill, as instructed by a computer in Houston, Texas. All the old records created by the pump operator and office clerk have disappeared—along with the operator and clerk, their co-workers and supervisors, the regional offices and their staffs, and all the folks in Houston who used to manage the people and information associated with this now-defunct data-collection apparatus. In their place a relatively mundane computer, a network of "phone" lines, some software, and remote input-output devices accomplish the entire job.

Here we see a clear-cut example of technology's power to simplify work through direct information links. Dozens of busy people employed one day, a virtually empty facility the next. And yet the operation continues to perform its function.

Understandably, many people will focus on the gloomy aspects of this story. "How can I prevent the same thing from happening to my job?" A good question. From my perspective, I'd worry about any job that merely transforms information from one form to another on its way to a computer. Such tasks seem sitting ducks for direct-access alternatives.

On the other hand, I hope that thoughtful readers will check this horizon for hints of opportunity as well as danger. Consider the Houston team in this story. A handful of people can now monitor transactions across an entire continent. What could such empowerment do for your business or, better yet, for a business that doesn't now exist? Whether as prudent risk avoiders or eager entrepreneurs, each of us can profit from a better understanding of the risks and opportunities offered by computer-aided access to needed information.

A generation ago, Stanley Kubrick's film *2001: A Space Odyssey* painted a gripping picture of a battle of wits between humans and a computer named HAL. In that struggle for dominance, an intrepid human ultimately managed to outwit his electronic opponent—but not before the latter succeeded in murdering most of the pesky humans who might have interfered with the computer's plans for accomplishing its mission.

In a real sense, the powerful competition between a computer and a human portrayed in this science fiction finds an echo in today's job market. Computers have had the last laugh, one might say—but today's machines still lack the intelligence to celebrate the victory that HAL once sought. Except for a few niche applications, the grandiose aspirations of Artificial Intelligence that HAL portrayed play no significant role in modern computer science. From a financial perspective, all significant gains from Artificial Intelligence (AI) have come from selling stock in promising start-ups to optimistic investors.

Though much of their energy and creativity now find outlets in mainstream computer science, more than a few AI enthusiasts still pursue their original interests—but the machines that dominate today's workplace have little pretension to intelligence. Diligent but dumb, automated teller machines quietly dispense cash, check bank balances, and print receipts. Slim pickings for science fiction, but enough functionality to revolutionize retail banking from New Delhi to Nome.

As in the fuel oil story I recounted, the ATM model offers an example of an apparently irreversible blow to sequential information handoffs—handoffs that constitute a long-standing feature of paper-based human information work. Now the simple act of keying in the amount of cash wanted by a customer bypasses the human chain that began at the teller's cage and ended with the keeper of the bank's general ledger. It also makes many traditional exchanges between banks unnecessary. In the past, for instance, an extended vacation or business trip might have required a series of letters of credit addressed to banks along the way. No more asking bank employees to write back and forth these days. We just look for ATM machines that accept our favorite flavor of plastic.

Today, a computer resides at the end of most information handoff chains. Think how many of our everyday errands result in changing stored strings of characters in some computer. Write a check to pay your gas bill, register for an evening course at your local community college, or trade some frequent-flier miles for an economy-class ticket to Barbados. Since the end result for each task consists of modifying appropriate entries in some computer, we can imagine an ATM-like simplification for each of these examples, and for thousands of others like them.

As with old-fashioned banks, each of these tasks presently involves help from layers of humans, most of whom perform tasks born in the paperwork era. Many of these intermediaries appear to add value—such as a knowledgeable travel agent who steers an economy-minded client to a low-priced charter package, for example. Someone who scans brochures and trade magazines for the best deals. Note, however, that much of this labor consists of sorting through eclectic information formats. Stacks of mail, directories, advertising of all sorts to be scanned, telephone calls to be placed and returned, and meetings to be attended. Lots of opportunities for the expenditure of time and effort on the client's behalf.

How much of this work will survive the coming transition from paper brochures to electronic formats? Imagine Home Shopping Network selling trips to Hawaii in an interactive setting much like a video game. While most consumers will welcome the prospect of more direct access to the information they seek, the white-collar workers who now reformat data in such arenas will need to find other ways of adding value. More direct information pathways promise to change traditional employment systems in drastic ways.

Risk and Opportunity

As computers get rid of paper along with the office workers who shuffle those papers, lifetime job security seems the relic of a bygone era. Today's explosive acquisition and application of knowledge has clearly thrust humankind into unfamiliar terrain, in which past experience provides little guidance. As investors, job seekers, parents, business managers—or merely concerned citizens—we can no longer safely guide our actions by the landmarks we once used. Where should we invest money, for example? Many blue-chip stocks with long track records have fallen upon really hard times. Today we not only see individual companies going bankrupt, but long-established industries like savings and loans and airlines that have wiped out much of their share owners' equity in the short span of a few months. But as we'll see shortly, other companies have prospered. In particular, many have found profitable opportunities at the peripheries of today's products and services as well as in the gaps between them.

What should job seekers look for in an employer, therefore? With almost half the U.S. workforce employed by companies that have downsized at least once in the past ten years, bear traps abound. Today, well-managed and highly profitable companies cut their workforce *before* they get into trouble, so prudent job seekers need new criteria to pinpoint

ies and avoid pitfalls. For example, I'd want to make sure that
ment in question had a broad understanding of its customers'
rnally focused work styles—the mind-set that produces paper-
r than new products—seem destined for trouble.

How can concerned parents prepare their children for the world
ahead? Technology and business practices are changing so rapidly that
job skills learned in school stand little chance of remaining relevant from
graduation until retirement. Clearly, the demands of a knowledge-intensive
economy will require preparation—but likely trends will favor certain
kinds of training over others. In particular, since we can expect accepted
bodies of knowledge to be overthrown multiple times in a single lifetime,
learning *how to learn* seems the most valuable skill.

What guidelines should managers adopt in crafting business strategies? The Information Revolution has apparently baffled more than a few
experienced CEOs. Some have floundered and lost their jobs, despite the
evident goodwill that earned them those positions and the high-priced
advice put at their disposal. Consider, therefore, the plight of managers at
lower levels. As I see it, managers at all levels would do better to join
moves toward direct information access rather than to engage in futile
rearguard actions against it.

Much of what we learned in the past still applies. Human nature
changes only slowly, after all. People still work, eat, sleep, make love, and
shop. Money, long-held grudges, and pop culture still hold sway in
human affairs. Even in the economic sector, traditional factors like labor
and capital, demand and supply, still matter.

But the accelerating acquisition and application of knowledge has dis-
turbed yesterday's equilibrium. Today, the onrush of modern technology
has thrust knowledge into such an important position that few areas of
human activity can afford to ignore its new role. The dominant power of
our age now flows from technology—the application of knowledge—and
examples of that power abound. We build a tunnel under the English
Channel, rearrange the genetic code of living beings, and light up entire
continents as seen from the spacecraft that circle our planet. Possession
of knowledge, it would seem, has put humankind in the driver's seat.

Moreover, as technology helps its possessors acquire ever-greater
amounts of knowledge, the pace keeps accelerating. According to
information-technology analyst Peter Keen, as much new text is now
published each day as all of civilization produced from the invention of

writing to the beginning of the nineteenth century. Microscopes allow humans to probe the surface of a virus on an atom-by-atom basis, while telescopes provide us with detailed chemical analyses of galaxies at the very edge of our universe. In between, telecommunications brings every corner of our planet as close as the nearest TV screen.

Most important, our kit of knowledge-building tools includes computers—machines that extend the reach and power of the human mind in much the same way that the invention of motors helped to extend the reach and power of human muscles. Taken together, all this new knowledge spurs us to create more technology, which helps us build more knowledge. Growing in an explosive relationship of its own, the interplay between knowledge and technology has now displaced labor and capital as the primary driver of the world's economy.

With less need for labor in agriculture and manufacturing, large numbers of workers in economically advanced countries find themselves sitting behind desks. At first glance, the eighteenth-century Luddites in England who saw textile machinery as a threat to human livelihood appear proven wrong. Machines do all the physical work, while humans handle the papers needed to keep things moving. But while paper abounds everywhere, paperwork—the unproductive tasks that organizations generate to meet internal needs—no longer seems indispensable. So, as we have already noted, many middle managers find themselves unneeded and unemployed.

At the same time, others find profitable ways of bridging information gaps that conventional wisdom has overlooked. For instance, a young businessman named Howard Tullman noted that the price list that auto insurance companies used to reimburse their policyholders for wrecked cars got somewhat stale between editions, and averaged out regional differences—such as the midwinter spread in the price of convertibles between Milwaukee and Miami, for instance. Seizing upon that opportunity, he started the Certified Collateral Corporation to refine this price list for the auto insurance companies.

Through a combination of traveling agents and telephone inquiries, Certified Collateral maintains an up-to-date inventory of the cars in virtually all U.S. used-car lots—year, model, make, options, and the price the owner in question will take for them. In practice, this drops the price of each car by something like $300 below that listed in the so-called Blue Book. With some one million "totaled" cars each year, that amounts to a tidy savings for the insurance companies that have to replace the totaled cars—even after CCC gets paid for its services.

Policyholders can take the money, a car just like the one they lost, or, quite frequently, a similar model a couple of years younger for an additional charge through a nearby dealer. The dealers like the arrangement because they sell extra cars, and they get help from CCC in keeping their inventory data up-to-date. Information has value, after all.

Alongside such bright spots, however, uncertainties abound. Economic theorists from Robert Reich to Peter Drucker cite information workers as the key resource of our age, yet employers around the world seem eager to dispense with their services in large numbers. Is this a cyclical downturn in the economy or a fundamental change in the nature of information work? Either way, it pays to learn as much as we can from what has happened so far.

The Winds of Change

Today, many of us have come to regard "revolutions" as everyday events. In the 1960s and 1970s, popular literature attached that label to the relatively tame introduction of computers and telecommunications into office work. Nothing more would happen, one might think, because every significant company had a full-blown computer center. Even the subsequent introduction of PCs into office work caused few ripples. The so-called Information Revolution had come to fruition, it seemed, somewhere in the 1980s.

But the end of revolution ought to bring a period of stability. Things ought to quiet down. Instead, the world economy finds itself in turmoil, with countries and companies beset by new forces. Traditional economics would predict job growth during periods of economic expansion. But corporate America's large-scale shedding of surplus staff began in the relatively prosperous 1980s, when the *Fortune* 500 companies reduced workforces by some four million employees. Chalk up that job loss to competitive pressures. During a time when sales and earnings climbed substantially for the group as a whole, major U.S. corporations made themselves more cost competitive by reining in their oversized payrolls.

Others adopted alternative strategies. Had Europe's 1992 unification strategy succeeded, multinational cooperation by European governments intent on nurturing domestic industries in key technologies—integrated circuits, computers, telecommunications, and consumer electronics—ought to have produced a formidable competitor to the United States and Japan. But Europe's unemployment rose into double-digit figures in the early 1990s. It might have gone even higher had the EEC nations not managed to export some of their unemployment by reversing the flow of temporary workers from their less fortunate neighbors to the south.

Japan's long-running record of economic success turned sour as well. Longtime winners such as Nissan Motors posted their first losses in over a generation. The Tokyo stock market plummeted, prompting massive government programs to bolster a sagging economy and to offset rising pressures on a shaky economic outlook. While Japan's economic outlook has since brightened, the lesson seems clear. Past success can't guarantee a trouble-free future.

As we look for clues to what lies in store for the world's economy, we can learn from the wave of change that has already swept its manufacturing sector. Advances in technology now permit us to produce a mind-boggling array of material goods for a fraction of the effort required in times past—goods that do more, work more reliably, and cost far less than their predecessors.

Consider the expensive-looking watches offered for sale on city sidewalks throughout the world. With a bit of bargaining, twenty or so dollars can get you a Rolex look-alike in Auckland, Amsterdam, or Albuquerque. Lots of people buy them. Some to fool their friends, I suppose, but mostly because the street vendors' offerings invariably keep accurate time until their batteries finally give out.

Beneath this quirky feature of the so-called underground economy lies a near miracle of manufacturing efficiency. Just work the economics backward. Since the vendors happily accept $30 and rarely go below $20, we can safely assume that the watches can't cost more than ten or so dollars from the middleman. Otherwise, the vendors and their suppliers wouldn't have enough room to make a profit. Subtracting the cost of the case, strap, face, crystal, and assembly leaves very few dollars to pay for the drive mechanism. And yet the producers of watch drives happily supply them at seemingly ridiculous prices.

What's going on? While the low-priced watches themselves generally come from garage-type operations in low-wage countries, the drives are made in sophisticated factories in places like Japan and Switzerland.

According to industry analysts, these factories produce drives for less money than the price of a single Sunday newspaper—all parts included. And look at that parts list. Battery, battery holder, calibrated quartz crystal, crystal oscillator circuit, counter, divider, amplifier, stepping motor, reduction gears, and whatever hardware it takes to connect the bits and pieces into a single unit.

Sounds impossible, you might say. But you can try a watch the next time you stroll downtown. As you make your selection, you'll probably pay close attention to the style and finish, but you can take the mechanism's accuracy for granted. Since each drive's total labor content (parts included) amounts to just a few minutes, the manufacturer couldn't afford to make them if even a small fraction of them needed repair or adjustment of any kind. The same drives sit in more expensive watches as well, but the extra frills that surround them obscure their low cost in this setting. Since they already work perfectly, more expensive drives make no economic sense. Moreover, lower-quality drives would cost *more* to produce, so one size fits all, as they say.

What resource underlies the watch-drive maker's remarkable success? Cheap labor? Creating each drive's individual components seems a daunting task for anyone, yet none costs more than pennies. Massive amounts of capital? Interest on assets would drive the price too high. That seems to leave improvement in the manufacturing process itself. The watch-drive producer's engineers combined what they knew about ultra-low-cost assembly with worldwide parts procurement and experimented until they achieved a goal, one that no combination of hard work and money could duplicate by itself.

Until recently, few, if any, office managers could boast of similar gains. In contrast to the benighted "rust belt" of popular imagination, therefore, modern manufacturers underwent revolutionary changes during the 1970s and 1980s—as we've already noted—well ahead of their paperwork-bound opposite numbers. Let's take a look and see what happened.

The First Information Revolution

For most of human history, the persistent need for more food and artifacts has tied economic success directly to the quantity of goods that a person or economic unit could produce. As increases in productivity brought greater abundance, however, that "quantity" era drew to a close, at least in economically developed countries. Today, with an overabundance of cars, TV sets, and fast foods competing with one another, *quantity* alone no longer defines productivity. Instead, quantity has given way to *quality* as the deciding issue.

As an engine of technology improvement, human ingratitude ranks with the best. When TV sets first appeared, eager consumers snapped them up. Manufacturers merely cranked out boxes as fast as they could. Later, of course, as the need for frequent trips to the repair shop became a deciding issue, higher-quality offerings drove the old-style sets from the market. As revolutionary change ushers in a new era, it sweeps away the old in the process. And as quality-minded producers gained market share, those still wedded to quantity-based notions lost out.

With hindsight, one can easily fault the losers' less-than-adequate attention to their customers' growing dissatisfaction. Producers accustomed to seeing their mission in terms of quantity remained mired in that obsolescent worldview and ignored signs that might have warned them of an emerging shift to quality. Winning the game under one set of rules left them ill-prepared to play under another. Quality, rather than quantity, had become the deciding factor in determining value.

As I see it, that shift from quantity to quality marked the end of one era and the dawn of another. Looked at in detail, the picture reveals fundamental changes, many as profound as those that took place when mass manufacture swept past cottage industry in the nineteenth century's Industrial Revolution.

QUANTITY ERA	QUALITY ERA
Mechanical technology	Programmable control
Economies of scale	Economies of speed
Advance-planning emphasis	Customer feedback emphasis
Hierarchical organizations	Team-based organizations
Value from volume	Value from performance
Technology islands	Technology overlaps
Environmental exploitation	Environmental concern

We will examine these items in subsequent chapters, but an early look at this list will help provide a framework for the discussion that follows. For now, just look how different life has become. Conventional wisdom accumulated over a lifetime lost its validity—much like the upheaval in established truths that took place during the Industrial Revolution. And, as before, survival demanded learning new ways of doing business.

But meaningful learning takes effort. America's manufacturing quality literature dates back to the 1930s. But despite the fact that books by early quality pioneers sat on the shelves of corporate libraries, the U.S. consumer electronics industry ceded virtually its entire market to overseas competitors—competitors who had learned the quality-improvement lessons contained in those books and who thereby produced higher-quality goods. Just as the ability to learn proved decisive in this wave of change, those best prepared to absorb new ideas and to act upon them now stand a better chance of adapting to the changes still ahead.

The Second Information Revolution?

Once again, we see pressures upon the status quo. Just look at the consumer electronics ads in our Sunday newspapers. There is so much to choose from, and all of it is so free from manufacturing defects that buyers rarely give product quality much thought when making their selections. In this environment, consumers seem more concerned with other issues: "Can I connect this videodisc player to my present TV set and still use my VCR?" "Who can I call if I can't figure out the instructions?"

And with good reason. As Donald Norman documents in *The Design of Everyday Things,* his insightful book about present-day artifacts, most designers pay more attention to plaudits from their colleagues than to the real needs of nonexpert users. Results range from personal anxiety at one's lack of "computer literacy" to spectacular system failures generally attributed to "human error." Even the familiar VCR presents drastic programming challenges to most of its owners.

Alongside this nagging dissonance between technology and its human users, we also see conflicts between technology and nature. While few thoughtful people would forgo technology's aid in shielding civilization from nature's vagaries, dreadful examples of environmental neglect abound.

Finally, technology seems at odds with itself. Separately conceived systems continually oblige the public to resolve conflicting demands. Just look at the appalling information disconnections in a typical hospital. With floor after floor crammed full of expensive equipment, its staffers scurry from one computer screen to the next with scribbled notes. Each machine, it seems, sits in a world of its own.

Despite enormous progress, modern technology still falls far short of fulfilling human needs. Taken together, the threefold task of harmonizing technology with people, with nature, and with itself offers us a challenging goal. The technologists' task seems tough, but why settle for less? Progress, after all, usually begins with dissatisfaction.

Such dissatisfaction signals a new shift in the value equation. Just as the move from quantity to quality brought opportunities to some and misfortune to others, early adapters to a new paradigm—what I describe as *harmony*—should find fertile ground for growth today, when there is increasing dissatisfaction with the status quo. For some, the move toward harmony may well prove a natural extension of the earlier journey to quality improvement. After all, if a VCR user can't operate a particular feature, that portion of the control might just as well be broken.

In today's management literature, we can already find individual factors fitting into such a harmony concept ("ease-of-use," "systems integration," "green manufacturing"). The need for a new paradigm seems to increase.

Today, the transition from quantity to quality still remains incomplete in some parts of the global business community—particularly those that depend upon government protection against competition. But the pace of change leaves little room for breathing space. With the elements of the quality era barely in place, a new set of parameters seems called for, driven by the merger of computing and communications technology that serves as a theme for much of tomorrow's technology:

QUALITY ERA	HARMONY ERA
Programmable control	Direct information access
Economies of speed	Economies of convenience
Customer feedback emphasis	Personalization emphasis
Team-based organizations	Architectural organizations
Value from performance	Value from coherence
Technology overlaps	Technology merger
Environmental concern	Environmental renewal

While change rarely makes people comfortable, standing still seems even worse. After all, who enjoys a business climate that has made "downsizing" a household world?

Preparing for the Future

As with earlier revolutions in agriculture and manufacturing, the new shift in value destroys as it builds. And in the present economy, millions of people still endure the negative consequences of change. Reports of trade deficits, layoffs, bankruptcies, and higher taxes fill the pages of our newspapers. While some economists characterize these circumstances as a cyclical ebb in the global economy, traditional fiscal stimuli—such as historically low interest rates—appear only marginally effective. With more goods of all kinds available than buyers for them, governments around the world work hard at keeping low-priced foreign goods out of their home markets in order to keep domestic producers in business. Technology has boosted humankind's ability to produce, some might say, beyond its ability to consume.

But I disagree. I can't imagine that today's offerings exhaust all possible means of making modern life more satisfying and convenient. Consider all the cards that most folks carry around. Some men I know take the statement "Just put this handy card in your wallet" so much to heart that they routinely put their wallets on the table before they sit down to a long meeting. Overstuffed wallets can even cause spinal distortions, a back specialist once told me.

Why should anyone have to carry more than one piece of identifying plastic? Sandwich in some electronics and program the card to perform multiple functions covering all your needs—banking, travel, shopping, telephone, library, parking access, clubs, and more. Instead of filling out insurance forms every time you visit the doctor, just swipe the card. It would contain emergency medical information and access codes to allow your physician to get copies of your X rays and treatment records—all under protective encryption to ensure your privacy.

Sound farfetched? With the technology in place, we can already see the first generation of *smart cards* entering the marketplace. While all these features lie within reach, I'm sure that a few years' experience will lead to a host of unforeseen applications—just as happened with ATMs, faxes, and similar amenities that pundits didn't anticipate. In one application I saw recently, a would-be traveler ordered airline tickets from her travel agent and had the flight information and other data downloaded onto her smart card over the phone. She later carried the card to a nearby ATM machine, slipped it in place, and received printed tickets.

The benefits can range from avoiding the hassle of a bulky wallet to saving a life with ready access to medical data. Most important, this new approach will provide increased personal control of personal information, thereby giving its individual users more satisfactory control of their personal affairs. With many such unexploited opportunities awaiting enterprising businesspeople, I see little reason for long-term despair. Today's rapid escalation of technological prowess challenges our ability to deal with changing circumstances. But as Branch Rickey of the Brooklyn Dodgers once said, "Luck occurs when opportunity meets preparation."

Whether you relish the prospect of leading the pack or merely wish to avoid missteps, careful preparation seems warranted. Where will tomorrow's most promising business opportunities lie? What changes will technology and future business needs bring about in the ways in which we organize our work and our prospects for employment? Which technologies promise the greatest economic benefit, and how can we assure our access to them? In this spirit, the following chapters will explore the interplay between key technologies and business trends to provide answers to these and other questions.

CHAPTER 2

COMPUTERS

From the introduction of computers into the workplace in the middle of the twentieth century to the merger of computing and communications in its final decade, the role of information technology in our economy has grown enormously. Over the past fifty years computers have changed the world. And they themselves have changed as well. As their capabilities have grown, human ingenuity has adapted these versatile tools to a variety of roles in carrying out a rapidly growing share of the world's work. So rapidly, some might say, that the pace of that change has outrun its supposed benefits.

Let's look at the interplay between computers and their users during the fifty or so years since these machines first reported for work. At the outset, computers helped a relatively small number of people, mainly scientists and engineers, shorten tedious hours of manual calculation. With computers everywhere today, information processing now constitutes a major portion of every economic activity in the world—from flying on an airplane to buying a pack of chewing gum at a neighborhood store.

Along the way, computers have made themselves indispensable by providing their users with economic benefits. But these benefits generally occur indirectly. Except for a few special items—video games, for instance—humans rarely consume information processing directly. You can't eat it, wear it, or live in it. Instead, computing conveys its benefits by improving the production of the goods and services that sustain our way of life.

While the advent of electronic computing might appear as an instantaneous event in human history, our closer-in vantage point renders it a more gradual process. Despite all the early hoopla about the impact of computerization, two generations of office workers cohabited with commercial computers of one sort or another before any major upheaval began. Over that time, computers took on a number of functions, each with its own impact upon human productivity.

The family of machines we call "computers" encompasses a rich variety of forms and functions. Today's computers range from the slivers of silicon that handle the insides of point-and-shoot cameras to the room-sized banks of blinking lights that sometimes appear as villains in science fiction movies. But all computers share the same family tree.

As I see it, this "tree" has sprouted five principal branches during the past half century—roughly one every decade. In the order of their arrival, they filled the following functional roles: *numerical calculation, data processing, monitoring and control, personal assistance,* and *communications enhancement.*

In tracing computing's impact upon productivity, this list of paradigms acts like the chapters in a novel. It divides a complex story into a sequence of more tractable pieces. In that spirit, let's examine each of these functional elements.

Numerical Calculation

Numerical calculation not only predated the advent of electronic computing, it also provided the original motive. Humankind's long-standing desire to improve its ability to "crunch numbers" fueled the process that ultimately led to today's Information Age. From real "digits" (i.e., fingers and toes) to "digital circuitry," people have supplemented their mental calculating capacity with computational tools. And from Blaise Pascal to John von Neumann, mathematicians have led the way in pioneering the mechanization of calculation. In 1646, Pascal designed a gear-based machine that enabled him to add numbers at the turn of a crank—some three hundred years before von Neumann completed the foundation for modern computing with his invention of the programmable architecture we know today.

With hindsight, we moderns can see a rich variety of computer applications extending far beyond number crunching, but the challenges presented by automating numerical calculation alone kept the early pioneers fully occupied. The need for large amounts of tedious computation extended to many aspects of life in the Industrial Age, well before electronics appeared on the scene. While business uses of arithmetic for accounting purposes date back into antiquity, more esoteric applications did not come into being until the advent of commercially significant scientific pursuits in the 1930s. On the other hand, other human activities, notably fundamental science and weaponry, shared a small but voracious need for "computers"—a job title given in those days to *human beings* who crunched numbers with hand-operated mechanical calculators. Their numbers never exceeded a few thousand worldwide, but such people filled an important role in the first half of the twentieth century.

While the average citizen remained unaware of "computers," a small but widespread need for number crunching predated the modern use of astrophysical models, smart bombs, and command-and-control centers. Consider the story of Karl Schwarzschild. An early contemporary of Albert Einstein, he laid the theoretical groundwork for modern stellar astrophysics. His legacy includes the application of Einstein's theory of relativity to the life history and evolution of stars—including the point-of-no-return (or Schwarzschild radius), which dooms the interiors of massive stars to inevitable collapse into black holes.

But Einstein kept only one of Schwarzschild's papers in his personal file—Schwarzschild's last, a mathematical treatise on the effects of atmospheric drag on artillery shells. Written in 1916, shortly after its author had spurned deferment in order to enlist in the German kaiser's army, it was published after his untimely death from a virulent infectious disease. The publishers evidently sent Einstein a copy, one that this life-long pacifist kept until his own death some forty years later.

In Schwarzschild's day, such calculations required many hours of tedious arithmetic. Even as late as World War II, physicists working on the Manhattan Project depended upon rooms full of *computers*—government employees who spent the day pushing buttons on desk calculators in order to provide numerical guidance for atomic-bomb designers.

With the advent of electronic computers, such manual efforts quickly gave way to mechanization. Today, even a lone scientist with a desktop workstation can outperform the first A-bomb's "computers." But today's nuclear-weapon designers don't confine themselves to desktop machines. Indeed, the U.S. Department of Energy's weapons laboratories employ the world's largest and most powerful supercomputers. Moreover, severe export restrictions prevent access to similar machines by would-be bomb designers in other countries—in much the same way that our government tries to prevent the spread of plutonium and long-range missiles.

While number-crunching computers have had little impact upon the modern workforce as a whole, numerical computation holds the key to sophisticated designs in much of science and engineering. In laboratories around the world, research physicists explore alternatives on a computer before trying out transistor designs for the next generation of ultracompact chips in the lab—just as aeronautics engineers evaluate airframe designs on their workstation screens long before making models for wind-tunnel testing.

Today, numerical calculation still serves both abstract scientific curiosity and the needs of national power. In particular, while supercomputer sales constitute only a fraction of 1 percent of Japan–U.S. trade, officials on both sides appear to regard the welfare of their respective supercomputer producers as a "life-and-death" matter. In one such case, the respective merits of the Cray III and Fujitsu's SX-3 appeared on the front pages of the *New York Times* and *Asahi Shimbun*. Should Japan's Fusion Science Institute have given more weight to Cray's superior ability to link multiple processors together via software than to Fujitsu's apparently faster hardware? The sale of a single machine—priced far lower than the cost of a wide-body airliner—brought forth "unattributed" hints of trade war.

While competition from far cheaper and more abundant microprocessors limits such machines to a tiny niche market, these finely tuned numerical processors afford their users significant aid in the kinds of calculations required to design a broad range of high-tech items, from supersonic aircraft to microscopic transistors. Furthermore, such specialized capabilities have also permitted great strides in calculation-intensive fields like weather forecasting. But they would be wasted on the task of figuring out who still owes money to the Internal Revenue Service. In this latter application, a number of users (IRS examiners, say) need to share access to enormous bodies of data as they search for—and then process—a relatively small number of largely unrelated items.

Historically speaking, this need to sort through massive records consumed far larger amounts of clerical labor than did the numerical problems described above. But the first computer scientists focused upon numerical calculation rather than data processing. No one thought of using computers for this latter application in those early days. As time passed, however, improvements in technology combined with application insights to form a still-flourishing second branch to our family tree, and with it the first significant impact of computers upon mainstream office work.

Data Processing

Data processing has grown into such an important part of computer use that many people have come to use the terms *data processing* and *computing* interchangeably. For at least twenty years, any distinction between them made little difference. The data-processing paradigm proved so successful that it dwarfed every other aspect of computing.

This dramatic growth in computing took its principal beneficiaries by surprise. Until the early 1960s, conventional wisdom saw the fledgling computer industry's primary role as displacing desk calculators. But the advent of affordable mass memory changed matters abruptly in 1954, when IBM introduced such a memory in its model 704. Among other features, it offered users one million bits of memory. While the $1-per-bit price tag seems staggeringly large by contemporary standards, the new memory represented a dramatic advance at the time. Computers cost millions anyway in those days, and the new memory supported a revolutionary expansion in applications.

Most important, the new memory facilitated an extension from numerals to the full range of keyboard symbols. Computer users could submit text-based tasks based upon high-level computing languages, such as FORTRAN, then coming into use. Moreover, data in the form of text—such as the names and addresses of Social Security recipients, heretofore stored on mechanically sorted punched cards—suited the new machinery just fine.

Instead of mere servants of R&D and accounting, these new *mainframes*, as they came to be called, appeared poised to rule the corporate roost. America's *Fortune* 500 companies—as well as most aspirants to that status—jumped at a technology that promised to automate their record keeping.

Fueled by that demand, a combination of technology advances and market opportunities led to a steady stream of system enhancements. Mainframes added specialized hardware to handle so-called peripherals—tape and disc drives for mass storage of data, printers, card readers, and, over time, multiple terminals—together with special facilities for moving data in and out of memory fast enough to keep the mainframe's central processor from wasting time waiting for work.

Did these mammoth machines make their printout-carrying human users more productive? Skeptics soon learned to confine their reservations to humorous asides. Any serious expression of doubt might label its originator as old-fashioned in an era of unquestioned belief in progress. Every indicator pointed upward, anyway. With the company as a whole making gobs of money, why nitpick about the productivity of any individual activity?

What a nice time to be in the computer business—especially for IBM, the market leader. Continued growth for vendor and customers seemed a foregone conclusion. Astute economists such as John Kenneth Galbraith heralded the advent of a "New Industrial State," one that would be dominated by virtually invulnerable giant corporations. Able to generate capital from their own operations, assured of customers through their marketing prowess, and protected from upstart competitors by economies of scale, these companies appeared destined to ride on an everlasting tide of progress.

In Galbraith's view, steady growth in revenue permitted corporate managers to grow their organizational hierarchies accordingly—and to thereby enjoy the benefits of ever-expanding career opportunities. As corporate planning "assured" continued expansion, it invariably called for more planners as well. The big new central computers helped the hierarchies to grow. As Galbraith observed: "Massive and complex business organizations are the tangible manifestation of advanced technology."

Despite these investments in centralized record keeping, the advent of mainframe computing barely touched the frontline aspects of business operations, such as manufacturing and sales. Industrial productivity still depended upon a foundation of mechanical technology, not significantly different from that of the early days of mass production. Major automobile manufacturers such as General Motors employed multidrill presses that could drill all the holes in an engine's cylinder head in a single step. A great saving, provided one needed enough engine heads to justify its design cost. Since any changes in that machine would cost huge sums, extensive up-front planning took place—with the concomitant expense chalked up to expected benefits from "economies of scale."

As the 1960s gave way to the 1970s, some chinks began to appear in the corporate armor. But few people suspected that America's corporate hierarchies—and their computers—would soon face stiff competition. Indeed, the assured growth theme survived intact in the fourth (and presumably final) edition of *The New Industrial State,* Professor Galbraith's epoch-making book published in the early 1980s. By that time, however, the success of overseas competitors and domestic start-ups provided growing evidence of competition.

In particular, a host of would-be competitors appeared in the seventies and eighties—none of whom seemed put off by the economies of scale that the still-confident incumbents depended upon. What had changed the picture so dramatically? While no single factor drove all the changes that began in the 1970s, it seems clear that the new computing technology offered intruders a new source of flexibility. And this flexibility, in turn, enabled these newcomers to circumvent many of the entrenched productivity advantages that incumbents had counted upon. In particular, many of the intruders took advantage of a new kind of computer and the new applications paradigm it made possible.

Monitoring and Control

As more and more people brought a growing variety of tasks to their local mainframes, regular trips to "the computer center" became a familiar feature of office work. By the mid-1960s, most manufacturers of office furniture offered optional drawer space for punched cards. Each computing task required a set of these cards for its input. Depending upon the size of a job, computer users either punched their own or submitted the task to a data-entry clerk.

Either way, each user's personal collection of punched cards maximized the computer's efficiency and minimized the amount of money that users had to pay for its services. But users paid in other ways. Each job submission and pickup normally took at least two pairs of round-trips, one for data entry and the other to the computer center itself. Moreover, minor mistakes often remained undetected until the final printout appeared. And that meant more trips and more waits.

Compared to the manual alternative of punching buttons on a desk calculator, computers generally saved these users a good bit of time—and also performed tasks too daunting for even the most diligent humans. But people rarely remain grateful for long. Once the tedious nature of manual computation faded into memory, only the day-to-day irritation with repeated waits remained.

As long as mainframes were the only game in town, however, users could do little more than double-check their work before submission, and perhaps cajole the local computer center into reducing turnaround times.

Matters improved when the advent of time-sharing introduced a more flexible form of interaction. In this arrangement, users could engage a computer's services via a terminal, thereby avoiding multiple trips to the computer center. The computer, for its part, accommodated multiple users by giving each task momentary attention in round-robin fashion. Thanks to the mainframe's blinding speed, most users could feel as if they had the computer's full-time attention—albeit with sharply reduced performance.

Despite occasional squabbles with users, computer center managers had much reason for self-satisfaction. Use continued to expand, machine power grew, and the importance of computing rose steadily in most large organizations.

The processing engines chugging away at the heart of each center's operation created their own economies of scale. For example, producing a large payroll required hundreds of operations for each one of thousands of employees. While a desk calculator typically took several seconds to grind its gears through the multiplication involved in computing a percentage for Social Security deduction, a mainframe might execute such a task in less than one-millionth of that time.

In order to achieve such levels of performance, processor designers resorted to the fastest electronic "gears" available and drove them as hard as the laws of physics allowed. Higher voltages produced larger currents in order to charge and discharge capacitors more rapidly. As ones and zeros flew in and out of each register, the physical heat generated challenged designers' ingenuity. Indeed, IBM was said to have spent something like $8 billion on the design and tooling for the cooling module for its top-of-the-line processor circuitry.

To speed things along, processor architects added specialized circuitry to execute compound instructions directly rather than requiring the software to sequence through a series of more primitive steps. Thirty-two-bit words—the equivalent of ten-digit decimal registers—became standard.

Complexity grew. Every time a logic designer added a new box to the layout, more transistors went in, and more heat had to come out. Moreover, each line drawn between elements translated into a bundle of thirty-two wires—one for each bit.

Again, increased speed and functionality produced further economies of scale and drew more tasks to this powerful central resource. "Do not fold, spindle, or mutilate" became a familiar admonition in millions of transactions. College students registering for classes, holders of U.S. savings bonds, citizens everywhere, were faced with the punched cardboard rectangles that mainframes swallowed so readily—once a particular task reached the head of the line.

Given the enormous investment a "computer" represented, waiting one's turn made perfect sense—until someone changed the rules. In December 1959, a young MIT-bred entrepreneur named Ken Olson and his colleagues introduced their PDP-1, the first of a new breed of computers. Unlike the new computers that IBM and the so-called BUNCH were bringing to market in those days, this machine didn't offer more power or extra functionality. In fact, it offered less on both counts than most other machines then on the market.

From our modern perspective, the notion of a small computer makes perfect sense. And in fact, Olson's new approach drove his fledgling Digital Equipment Company past every established competitor except IBM itself—clearly demonstrating the novelty of this new computing paradigm.

At the time, however, DEC's first machine seemed more suited to filling a tiny niche than to changing the definition of computing. It consisted of just a pair of metal cabinets filled with low-tech components. In contrast to the costly sophistication of its mainframe cousins, the PDP-1 seemed almost homemade—more like a hobbyist's creation than a serious commercial entry.

But Olson and his colleagues were *very* serious. DEC's PDP-1 represented a thoughtful trade-off between functionality and cost. They stripped the former to the bone in order to bring the latter within the reach of a mass market.

First of all, the PDP-1 handled only eight bits rather than the mainframe's thirty-two. While this smaller word size required at least four instruction cycles to replicate a single beat in a mainframe for some applications, it slashed wiring complexity by 75 percent. Similarly, the design jettisoned specialized hardware shortcuts in favor of multistep work-arounds written in software.

The machine didn't just take more steps to get through a problem, it also took more time per step, owing to its slower circuitry. As a result, PDP-1 buyers generally got less computing per dollar than a mainframe might have offered them. But owning their own machines eliminated the need to wait in line. And that made all the difference. Instead of bringing tasks to a computer, humans could now bring computers to specific tasks—such as controlling machinery and monitoring its behavior.

As time went on, the availability of supervision by computers instead of people allowed system designers to impose coordinated electronic control upon each step of a sequential process—notably in manufacturing. Indeed, even though offices took on the lion's share of computing machinery in the quarter century following the advent of *minicomputers*, as they came to be called, factories appear to have gained the greater benefit. The fact that manufacturing productivity tripled during this period, while office productivity barely budged, lends support to this conclusion.

Let's look at how the advent of electronic process control changed manufacturing. In the mechanically-driven regime preceding it, factory designers confronted a trade-off between flexible (but costly) human labor and inflexible (and even more costly) machinery. At large volumes, the up-front costs of complex machines—like General Motors' multibit drill presses, for instance—generally paid for themselves. And, as we have seen, their associated economies of scale kept upstart competitors at bay in most industries.

With computers controlling a manufacturing line, changes became simpler. The range of tasks performed by individual machines began to broaden, and specialized tooling costs declined. As these trends combined to make smaller lot sizes affordable, they eased the path of new entrants into the ball game. But entry alone doesn't guarantee a win. Why switch to an untried product just because it costs a bit less?

Alongside these cost factors, however, dedicated computer technology led to another benefit that redefined "productivity." Until then, mainstream thinkers saw "productivity" in terms of quantity—the number of cars, can openers, or cameras a given-sized factory could produce each day. With computer control, that broadened to encompass an additional dimension—quality.

While each mechanical solution stands essentially alone, data-driven machines all share the same language of ones and zeros. And this commonality not only aided overall supervision, it also allowed adjacent modules to swap information—and thereby identify interface problems as they occurred. In the mechanical world, on the other hand, any need for correction generally surfaced only at the final, end-of-the-line inspection. The rest is history. As quality improved, customers abandoned the old-style producers in droves.

Over time, the advent of minicomputers not only led to process improvement, it also set the stage for the next computing paradigm. If a group of people—or a buildingful of machinery—could command the full-time services of a dedicated computer, might something even smaller provide one-on-one services? As you probably know, the answer arrived in the early 1970s. Again, a new group of players led the way.

Personal Assistance

In 1970, Xerox established its enormously productive Palo Alto Research Center. Among other accomplishments, the self-proclaimed "Xeroids" who worked there created the Ethernet link between computers, an early computer mouse, the first "windows," and much else. But their "Alto" outshone all the others. Expensive and complex, it nonetheless embodied the functions and attributes we now recognize as that everyday appliance, the personal computer.*

Except for the laser printer, which this same group also produced, the parent corporation experienced considerable difficulty in bringing PARC's advances to market in a timely way. Still, the advances themselves drew admiring attention from technophiles the world over. In particular, a young enthusiast named Steve Jobs visited Xerox Park and liked what he saw. And soon thereafter, Jobs and his partner, Steve Wozniak, embarked on a new venture—a do-it-yourself kit that enabled electronic hobbyists to wire up a small but fully functional computer for themselves.

From that modest beginning, the company that Wozniak and Jobs had named Apple soon sprouted factory-wired models. As their success grew, other companies rushed into this new market. Finally, IBM itself entered

---

*Strictly speaking, the credit for inventing the first personal computer actually goes to a Texas-based team of engineers whose smart terminal operated as a stand-alone machine early in 1971.

the game as well. In order to encourage independent software vendors to write applications for their new PCs, IBM created a de facto standard by anointing a small company's disk operating system, or DOS, as it came to be called. The "small company" didn't remain small for long, of course. Today the total value of Microsoft's shares rivals that of IBM's.

With IBM's endorsement, PCs and their applications gained increased business acceptance. But, while anecdotal evidence and personal observation attest to individual benefits from spreadsheets and word-processing programs, U.S. government productivity figures failed to show significant gains in office-work output during the 1970s or 1980s. Despite heavy business investment in personal computers, offices clearly realized far fewer benefits than factories did.

I speculate that early attempts to "computerize" offices lost as much ground as they gained by adding unproductive rigidity to poorly understood procedures. In other words, experienced humans generally bend a rule or two to get the job done, and computerization generally makes this less possible. While I can't prove direct causality, I observe that moves toward individual empowerment surfaced around 1990, at the same time that "white-collar" productivity finally began to increase.

At least through the late 1980s, however, "personal" computers for factories showed clearer yields than did office-based ones. As machines performed well-defined tasks under microprocessor control, their controllers could improve performance by monitoring results and making appropriate adjustments based upon clear-cut data. For example, the parts list, required assembly steps, and performance criteria for a fully functional VCR provide such objective benchmarks for process control.

Since microprocessors "speak," and respond to, a common set of languages, communication and coordination among machines becomes easier. And all of this improves quality. By the early 1980s it had dawned on most consumers that the products offered by new entrants (notably from Japan) worked more reliably than the offerings of long-established firms. As a result, U.S. companies' market shares dropped substantially in one industry after another. After some false starts, however, "quality" became a piety in all but the most benighted of U.S. businesses, and, over time, the quality lessons learned on factory floors began to impact upon office work as well.

How do the principles of quality-oriented manufacturing apply to office work? In old-fashioned factories, each machine operated according to a rigid, preset plan, without communicating with its neighbors. Hierarchically organized bureaucracies operated in much the same fashion, and stand-alone computers did little to change the situation. Like a modern factory, on the other hand, present-day knowledge work depends upon a two-way information exchange that extends from a product's conception to the end of its life in a recycling bin. In that endeavor, the people involved must share ambiguities as well as so-called hard data. Fortunately, machines based upon rigid zeros and ones can play significant roles in this new arena as well, especially in functioning under yet another computing paradigm, one just beginning to emerge.

Communications Enhancement

While computers became involved in communications as early as the 1950s, steady increases in technological power have combined with a growing body of innovation to produce a new entity—one that provides its users with powerful links to people, to machines, and to data. Together, these three application areas promise more productivity gains than any we have yet seen.

From an individual user's perspective, the most evident product of this new paradigm will emerge from interpersonal communications—notably multimedia teleconferencing. The display on one's palm-sized pocket computer, electronic notebook, desktop PC, or conference-room projection screen might contain half a dozen video windows, each bringing remote conferees' voices and images from their respective locations. Moreover, shared software, graphics, and documents create a common visual space in which all participants can interact with subject matter in full view.

In such an environment, participants may enter and exit conferences as needs indicate, rather than remain locked in or out by the need to maintain a physical presence. Moreover, conferences may continue even after all participants leave in some cases, permitting later reentry for review and continuation of the matters under consideration. As a result, this conferencing style will not only support participation from individuals at multiple locations, it will also enable individuals to monitor, and participate in, multiple meetings as well. Much less wasted time, and more opportunities for productive impact.

Similarly, in the area of communications between people and machines, we can now see rapid progress beyond mere keyboarding. In particular, "computers you can talk to" appear slated for common use by the late 1990s. One of my Bell Labs colleagues used to keep a speech-recognition system—microphone, loudspeaker, and computer—ready in a corner in his lab for the steady stream of visitors who wanted a demonstration. Nowadays, he doesn't bother. Instead, he just flicks on his speakerphone, dials 0 plus a friend's number, and pretends that his phone doesn't have Touch-Tone keys. When the "operator's" recorded voice asks him to say "collect, third party, or calling card," he merely speaks his reply in order to put the recognition software he invented through its paces for the visitor's benefit.

In addition to speech, today's newest computers also recognize scribbles. While relatively few pen-based electronic tablets reached individual users in the early 1990s, commercial uses began to blossom during that period. Many package delivery services collect electronic signatures via electronic tablets, stockbrokers use similar devices to record trades, and warehouse personnel jot down information as they walk down aisles taking inventory.

While most of us prefer to pace our information intake by reading neatly typed text, we also generally prefer to speak or scribble rather than type. When someone calls me with anything more than a short question, I usually try to get them to put the matter on paper. (Unless, of course, it's my boss. *He* frequently asks me to write down my ideas so that he can examine them more carefully.) Since we deserve to act as "bosses" to our computers, speech and handwriting recognition capabilities will allow humans to communicate in their preferred styles.

Finally, this new computing paradigm promises to make its users more productive by offering enhanced access to data—mining large bodies of raw data for tiny nuggets of information and bringing the results into visualizable focus. For example, an individual computer user connected to the Internet can now have access to the card catalog of the Library of Congress—as well as to hundreds of other catalogs around the world.

While the data-processing paradigm enables access to individual databases, a new computing style based upon large numbers of processors working in tandem has enriched this application dramatically. In particular, the huge increase in processing power of the new machine enables its users to sift through multiple databases to bring hidden facts into focus. Consider the kinds of information that might aid in tracking infectious diseases, for example. Does traveling south in midwinter increase one's chances of catching cold? Yes, one could get an answer by hiring interviewers to carry out the legwork, but let's see what a single computer-aided person might accomplish by searching through existing databases.

Since most travelers use credit cards, the card companies' billing records could provide our hypothetical searcher with a list of people who visited various cities, along with an outline of what each one did during his or her stay. Back home, insurance claim forms would come into play if the traveler in question gets sick enough to see a doctor. Did the weather play a role? Each lookup generates multiple searches.

Until recently, the amount of computation demanded by such tasks exceeded generally available computing resources. But that changed with the advent of multiprocessor computers. Indeed, the most powerful computers in private hands during the early 1990s devoted most of their cycles to database applications. As possibilities for more sophisticated data mining continue to grow, the value of the information gained from database comparisons will place ever-greater demands on our newest computing paradigm. How will analysts cope with masses of such data, in this and other fields? Here, too, we can expect this new computer paradigm to enhance knowledge, especially by enabling visualization.

Think of visualization as the ultimate spreadsheet. Like a child's first encounter with a pile of nesting blocks, users can manipulate three-dimensional images and "feel" them at a nonverbal, almost preconscious, level. "Playing with the data," as it were, but in a deeper sense than heretofore possible.

As these applications enhance productivity by building links to other people, to machines of all kinds, and to the world at large, the new computing paradigm can help bring about a benign resolution of our present-day economic dilemma: what to do with the "surplus" labor force created by recent jumps in productivity. Since new opportunities appear most plentiful at the interfaces of existing businesses and in the gaps between them, most of us will need to expand our horizons. Here the merger of computing and communication can help by providing new means of linking up with customers and new ways to serve their needs. In our society those needs assert themselves through a mechanism we call the marketplace—a subject we will explore in the next chapter.

CHAPTER 3

COMPETITION

When I came to America as a young child, my family crossed the Atlantic on one of the proud fleet of ocean liners built along Scotland's River Clyde. Today, those shipyards lie rusting and empty—the same fate that later overtook the Clyde's successors, shipyards in such countries as the United States and Sweden. But the story doesn't end there. In the latter half of 1993, Japan lost its leadership as the world's largest shipbuilder to Korea, a country with far fewer people, a small fraction of the former's economic prowess, and until recently no significant role in this industry.

What a contrast to the economic certainty that marked the early decades of the postwar era. If anything, most economic theorists worried about the effects of too much corporate staying power rather than too little. Remember, for example, that the U.S. government's attempt to break up IBM persisted into the early 1980s. Times have certainly changed.

As we saw in the last chapter, many of yesterday's most efficient producers have since lost out to others. But competition doesn't stop once the lead changes. After all, customers can simply go elsewhere whenever better values appear. So grueling, head-to-head competition continues. And, just as yesterday's winners found themselves outdistanced by newcomers, a present-day lead doesn't guarantee permanent advantage. Few safe harbors exist in today's competitive environment.

Most of us, I venture to say, regard competition with mixed feelings—especially when someone else gains at our expense. No one enjoys losing, after all. But much as we might sometimes deplore its consequences, our actions as customers encourage it. As citizens, we expect our elected leaders to make our nation economically competitive; as investors, we demand maximum earnings from our stocks and savings; and, as consumers, we reward efficiency by "shopping around" for the best combination of price and quality. Customers spur competition. And as technology continues to enhance customer power, we can expect competition to stiffen even further.

In this context, we need not limit the notion of "customers" to people with cash or credit card in hand. Instead, this book's broader use of the term follows modern quality-improvement terminology. In any sequence of processes, each process plays the role of "customer" to its upstream neighbor and, by extension, to that one's upstream neighbor as well. Customer feedback, then, becomes the first step in competitively driven process improvement.

Herein lies a paradox. While we seek economic security for ourselves, the sum total of our actions as customers drives change in the status quo. Over time, the balance of power between two forces shifts back and forth. Today, of course, competition sits firmly in the driver's seat.

Monopolies, West and East

In the first wave of the Information Revolution, the 1970s marked a period of transition. Until then, America's industrial giants stood comfortably at the peak of worldwide economic dominance. With meaningful competition relegated to the past, corporate congeniality ruled the day.

When selecting executive trainees in the 1950s and 1960s, most corporations sought cooperative—rather than competitive—candidates. In describing that era in *The Organization Man*, William H. Whyte began with an account of his own experience with a training program at Vick's, a corporation then still controlled by the owner-entrepreneur of the same name. Mr. Vick, it seems, valued competitively minded managers and saw his program as a contest among would-be warriors for commissions in a competitive army. So Whyte described this example only to disown it as a relic of the past.

In Whyte's day, executive training programs shunned competition among candidates. Everyone cooperated. The greatest success went to those who managed to cooperate better than the rest. Cooperation in that clubby era meant the ability to get along with like-minded colleagues. Hardly a forum for anticipating disruptive changes. If a bureaucrat thought at all about customers, it was to modify *their* behavior to suit his needs rather than the other way around. Who would want to rock such a comfortable boat, after all?

Reflecting that environment, a widely read 1950s novel depicted a fictional executive faced with a career problem. Should he struggle toward the top or just settle for the "limited dream"? The latter option offered a satisfying job, an assured series of steady promotions in a growing organization, and enough leisure time to enjoy the home, family, and social amenities that came with them. Nevertheless, would the lure of the "fast track to the top" entice this prototypical *Man in the Gray Flannel Suit* into working longer hours and agreeing to out-of-town trips?

In those days, it seems, career options matched the rating system used by the Sears Roebuck catalog of the era: "Good," "Better," and "Best." Everyone won. All one had to do was to decide on the amount. With Mr. Vick's view of competition seemingly relegated to the past, this fictional character stood as the hallmark of an age whose downside risks seem trivial by the standards of today's business environment.

What could go wrong? Even the most perceptive of post–World War II economists celebrated the victory of massive corporations over the vagaries of uncontrolled competition. Since such organizations could look forward to a future of uninterrupted growth, their managers found ample reason to anticipate fortunate futures—much like the inhabitants of a well-situated stone fortress in the days before gunpowder.

At midcentury, the awesome power of America's industrial incumbency seemed self-evident. By far the world leader in a host of well-established industries—automobiles, steel, pharmaceuticals, telecommunications, aircraft, electronics—Americans had every reason to expect comparable success in new areas then being spawned by U.S.-dominated innovation, such as the transistor, electronic computers, and nuclear energy.

Who could match the enormous economies of scale enjoyed by companies like General Motors and RCA? Besides enjoying the benefits of mass production, each commanded the loyalty of tens of millions of customers, and each deployed extensive dealer networks ready with spare parts and trained repair people in every community. How could any upstart provide comparable offerings?

But then the value paradigm shifted abruptly. As *quality* became a requirement for consumer acceptance, competition reappeared, and a flood of new entrants displaced long-established leaders in industry after industry. Take consumer electronics. The household names that dominated this field in the days of vacuum tubes survive as shadows of their former selves, if at all. Of this group, only Motorola managed to find a place for itself among today's leading manufacturers of electronic components—and it had to shed its traditional consumer businesses in order to stay the course.

With corporations no longer invulnerable, can governments protect their citizens from the pressures of foreign competition? Most governments rely upon tariffs, trade agreements, and the like to regulate the flow of goods and services across their borders. But the growing information content in economic values of all kinds makes such regulation increasingly difficult. (Interested readers can find a compelling account of such difficulties in Walter Wriston's book, *Beyond Sovereignty*.) Furthermore, some economic theorists see trade barriers as counterproductive, suggesting that the advent of a global economy should have us encourage the most efficient producer, irrespective of that producer's location.

But theorists rarely achieve unanimity. While some see competition as inevitable, others see it as a gratuitous waste of resources. Until recently, this latter notion dominated economic thought in much of the world—most notably in the planned economies that stretched from East Berlin to Hanoi.

Born in the aftermath of the First World War, the Soviet Union survived one of history's bloodiest invasions in the Second. Foreign occupation and civil war marked its early years. Hardened by cruel dictatorship and horrendous famines, the Soviet people overcame one obstacle after another to enter the latter half of the twentieth century as the world's second superpower, poised—in the eyes of many—to dominate all of humankind.

No other nation in history has aligned itself more closely with technology. "Communism Is Electrification," a typical slogan proclaimed. Steel cities, like Stalingrad, rose as cathedrals had in earlier times, while a steady succession of Five Year Plans strode across the calendar.

Steel, petroleum, railroads. By any measure, the Soviet Union's industrial base matched any the world had ever known. Larger in land area and population and richer in natural resources than its great rival in the West, it also boasted more engineers, teachers, and scientists. Moreover, these impressive achievements reflected quality as well as quantity. Soviet preeminence in mathematics and theoretical physics demonstrated itself in that nation's rapid creation of nuclear weapons.

With a disheartened West struggling to match its displays of raw power, who could have predicted that this last of history's great empires would falter toward collapse a scant thirty years later? Seemingly ready to take on the entire world in a nuclear war, if necessary, the leaders of this monolithic state found the base of their power crumbling. Over four generations its citizens had bled to invest in future economic growth, yet the Soviet economy lay in shambles. Bureaucrats still "made" their numbers, but this "official" progress contrasted sharply with the unmet needs of an increasingly desperate public.

By the mid-1980s, the citizens who had once taken pride in Sputnik now recognized their products as nothing more than shoddy imitations of superior Western models. And the sum of these quality shortcomings ultimately doomed the entire system to failure. Why? As we saw earlier, technology enabled a shift in industrial production—from rigid mechanical control to networked computer management. While the rigid inflexibility characterizing mechanical systems lends itself to monolithic top-down planning and control, computer networks excel at moving information sideways, from one end of the production process to the other. Imagine factory engineers exchanging e-mail with their customers under a political system in which even copying machines must stay under lock and key.

Down to their last futile *coup d'état*, however, the Soviet Union's traditional leaders sought to maintain the power they derived from controlling information flow. In such an environment, electronic bulletin boards, desktop publishing, and the other technology tools for horizontal communication threaten the system. Communication between adjacent processes had to move first up one line in the hierarchy and then down another instead of flowing freely among customers and suppliers. And for good reason. Real customers derive their power from an ability to make choices—and choice implies competition.

Economic success in the modern world, it seems to me, favors systems that take direction from their customers. Henry Ford's "They can have my cars in any color they want, as long as it's black" might well have come from the mouth of a Soviet bureaucrat rather than from the twentieth century's most celebrated capitalist. Either way, however, the top-down attitude that these words reflect now seems permanently ensconced in the dustbin of history. As a result, today's competitive drive toward customer orientation can no longer sustain yesterday's bureaucracies. The hordes of office workers who performed internally focused tasks in the monopoly-dominated era will inevitably have to find more productive ways to make a living.

Paperwork

In offices around the world, a growing number of companies have begun to apply the lessons learned on factory floors to the flood of paperwork occupying so much of their workforce. Competitive pressures confront them with a simple choice. Streamline the office, or go out of business. While factories continue to improve, ongoing productivity improvements have so reduced direct labor costs that relatively few dollars remain to be mined. Industry norms indicate that the percentage of a large computer's selling price actually attributable to the factory that assembles it has now dropped into the single digits. No wonder that most major computer companies announce major cutbacks in office staffs when they get into financial trouble these days.

Interestingly, most large businesses appear able to cut back their office staff without significant impact on their operations, and with significant benefit to the bottom line. While we all deplore the negative impacts upon displaced human beings, the positive economic results achieved suggest that the paperwork in question contributed a net negative value.

Herein lies the issue. Hierarchies helped manage complexity—and once played a vital role when large-scale mechanical manufacturing technology demanded vast amounts of up-front planning and top-down coordination. But such structures reduce flexibility as well as cost money—especially as paperwork takes on a life of its own. No wonder that competitively minded managements strive to shed such encumbrances whenever possible.

In addition to exacting a monetary burden, hierarchical bureaucracies impeded their owners' ability to compete effectively by slowing the pace of decision making. Traditional layers of management move information up and down a pyramid—abstracting it on the way up and elaborating upon it on the way down. At every layer, another level of managers spends time making decisions. Unimpeded transmittal would leave them with nothing to do. In a bureaucratic environment, therefore, a major decision's round-trip time from the front line to top management, and back again, frequently takes months.

Even worse, information filtering often reflects the interests of the organization rather than the needs of the business. Imagine a traditionally organized major computer company. In addition to its established mainframe, minicomputer, and personal computer businesses, let's suppose that this company recently began work on a line of workstations. How might the hierarchy influence the company's entry into this new market?

Suppose the company's management committee received a request for resources from its fledgling workstation business. In the normal course of events, the senior officers on that committee would send this request down their respective pyramids for comment. Top managers can't know as much as the experts who work for them, after all. But how do these "experts" react? To some extent, each existing product line sees the new entry as a competitor—for corporate resources and for market share.

Everyone looks busy. Meetings take place. Interorganization task forces study the issues and prepare voluminous reports. As the bureaucracy defends itself against its internal competition, more nimble single-product companies use this time to establish themselves in the marketplace. Whatever value such corporate processes may have had in the past, the real-life outcome sends a clear message. Upstarts like Sun Microsystems and Silicon Graphics grew to billion-dollar size by selling workstations, while the internally focused hierarchies in established companies spent most of their energy impeding the entry of a cross-elastic product. Even high-technology companies, it seems, have trouble with counterproductive paperworkers.

Does massive paper shuffling signal the fruition of the Information Revolution or the last gasp of the mechanically driven industrial era that preceded it? Based on our earlier discussion we saw bureaucracy comfortably ensconced in the Quantity Era, and under increasing competitive pressure in the Quality Era that followed. In one industry after another—televisions, copiers, automobiles, cameras, and a dozen others—entrenched incumbents lost out in head-to-head competition against vendors who offered better, and generally cheaper, versions of those same products. As a matter of simple survival, therefore, the defenders improved the quality of their offerings and trimmed their staffs to reduce costs.

But merely cutting back on paperwork may not prove sufficient. As we look to the future, I speculate that offices designed to filter information in hierarchical fashion will soon find themselves eclipsed by more nimble competitors organized for direct information access. The case of Sam Walton provides a fine example. In creating Wal★Mart, his phenomenally successful retail enterprise, he replaced much of traditional decision making with direct information sharing at the working level. Instead of attempting incremental improvements in its purchasing/receiving/accounts-payable systems, for instance, Wal★Mart teamed with one of its suppliers to create a radically different process.

Anyone who has cared for small children since the introduction of disposable diapers knows the importance of this particular item. Except for milk or infant formula, no household item exerts as much power to send someone to the store at any hour of the day or night. Unless they don't care about disappointing their customers, retailers pay particular attention to keeping an adequate diaper supply on their shelves—as well as in the warehouses backing them up. Much like concerned parents, retailers place hurry-up calls to their suppliers when stocks run low.

Until Wal★Mart's initiative changed the rules of the game, diaper manufacturers managed as best they could to match their production lines to the fits and starts of their customers' orders. Production engineers like to keep a steady flow in their lines. Stops and starts don't just annoy them, they eat deeply into earnings.

In 1991, therefore, Walton approached Procter & Gamble with a different proposition. P&G would no longer wait until some purchasing agent decided to place a diaper order. Instead, P&G would take over responsibility for keeping shelves filled and would get paid automatically as that product flowed past Wal★Mart's checkout scanners. By linking manufacture to up-to-the-minute sales data—instead of to periodic large orders from the purchasing organization—P&G smoothed out its production and also reduced the size of the inventory it had maintained to guard against uncertainty.

The success of this pilot program opened the way for its extension to P&G's entire product line, and later to other suppliers as well. Furthermore, the system has since taken hold with other leading-edge retailers anxious to keep up with the competition. Important-looking documents no longer flow in large quantities across interorganizational boundaries, and no longer provide internally focused work for the organizations involved—or even justification for their continued existence in some cases. In their place, a deceptively simple concept provides that regular payments are sent to suppliers automatically—almost as if they were part of the retail operation they supply.

But information sharing hasn't just slashed paperwork costs. By 1993, for example, Wal★Mart and its Kmart rival deployed the two most powerful computers then in private hands to manage their respective databases. With some two thousand stores apiece and over eighty thousand individual items per store, keeping track of prices, sales, inventory, and related matters on an item-by-item basis resulted in approximately one billion active entries into those computers.

With all information accessible to retailer and suppliers in a timely way—together with detailed, day-by-day data extending back several months—these two merchandising giants refocused their efforts from paper shuffling to strategy. Instead of matching invoices and manifests against purchase orders and store inventories, managers now spend their time going over detailed product-by-product marketing plans with each of their suppliers. Small wonder that their competitors struggle to keep their own stores afloat. While most retailers content themselves with six inventory turns per year, a 1992 *Business Week* article reported that some Wal★Mart departments managed as many as sixty.

The Wal★Mart chain now structures much of its business around the information captured in every swipe across cash-register scanners. In particular, scanner data flows into a corporate database serving everyone on the team. A houseware-section manager in Park City, Utah, and a senior vice-president of marketing in the home office can both track sales trends as they unfold instead of waiting for reports to appear after the fact. In addition to conserving financial resources that would otherwise be tied up in inventory, Wal★Mart's information sharing simplifies its work flow as well. Buyers and soap salesmen no longer hold meetings to predict what kind of detergent suburbanites will find most appealing. They let the customers themselves decide.

By intervening between retail customers and wholesale suppliers only where they can add value, Wal★Mart's employees generate more profit per hour than their opposite numbers at competing chains. Despite industry norms suggesting that mass marketers can expect no more than half the profits per employee generated in department stores, Wal★Mart outperformed Sears by almost two to one in this category. Small wonder that Sears cut back the number of its stores while Wal★Mart has continued to expand.

As the advantages of information sharing yield a decisive competitive edge in one industry after another, it seems clear that traditional office work faces the same kind of upheaval that was earlier experienced on the assembly line. With microprocessors everywhere, does labor-intensive paper shuffling make any more sense than old-fashioned factories once did? From that perspective, yesterday's office environment becomes a point of departure toward a rapidly changing future.

How one uses paper, rather than that medium itself, has become the competitive issue. Still mired in internally focused paperwork, too many of today's harried knowledge workers scurry from one meeting to another and pick up phone-message slips on the fly. One wonders how much net value such activities actually contribute. Today, the traditional office seems destined for change. A small but growing number of people carry their "offices" with them. The widespread acceptance of car telephones, beepers, faxes, and laptop computers acts as fuel for further innovation and spurs even more widespread use. As mobility increases, traditional office-work strategies no longer apply.

On the theory that it hired salespeople to talk to customers, Compaq closed its sales offices and instructed its salespeople to work out of their homes instead. To support this home-office arrangement, the company gave each of its salespeople a fully equipped computer and networked access to comprehensive databases. In that way, each salesperson can now review complete and up-to-date corporate data on a given client moments before making each sales call.

The result? With a sales force just two-thirds its earlier size, Compaq managed to double revenue while cutting unit prices in half. All in just a bit more than two years. With fewer internally generated demands upon their time and direct access to needed information, each Compaq salesperson sold six times as many computers as under the old system.

What will happen next? The impact of the head-to-head conflicts touched upon earlier hasn't reached everyone yet, and more belt-tightening appears on the way—especially in the internally focused aspects of "white-collar" employment we call "paperwork." But the single-minded focus on becoming leaner and meaner than one's competitor leads to nothing more than an exercise in pain endurance. Moving information more cheaply so as to reduce the cost of existing products remains a competitive necessity but hardly qualifies as a long-term strategy.

Information Value

As global competition increases, extracting maximum value from available information becomes a vital necessity. And in turn, competitive advantage shifts from the traditional triad of labor, capital, and material resources to knowledge and technology. Just look at the thirty-year-old billionaires who own companies like Microsoft and Dell Computers.

But how does a grandiose concept like "knowledge" help one to choose a winning strategy? In the previous chapter I described the limitations of mechanically driven manufacture—as exemplified by a multibit boring machine for automobile engines. Each attribute of such a system reflected the state of its designers' knowledge at the moment of a design decision. Here, knowledge lies frozen in time. Contrast this with a production process in which individual elements can exchange information about one another's needs and circumstances. As each element adapts to its changing view of downstream requirements and upstream resources, the entire process benefits from maximum use of available knowledge.

Some may view knowledge as an earned attribute that only increases—like the number of someone's birthday candles, for instance. Our definition, on the other hand, allows for ebb and flow. As "belief about the state of the world and its workings," knowledge is a perishable commodity in a rapidly changing world. As technology puts unprecedented power into human hands, *knowing what to do* becomes the paramount element of competitive advantage. This holds true not only in the factory and in office work but in vendor–customer relationships as well.

Walter Wriston, who moved Citibank past its larger rivals to the top spot in the U.S. banking industry, ranks as one of modern business's most accomplished and farsighted leaders. In the early 1980s Wriston began warning his colleagues of the coming revolution in the value of information. In Wriston's view, a bank stood to make more money from knowing what ought to be done than from actually doing it. This was such an unusual idea that few people understood it at the time. But let's take an example.

Imagine two telephone companies. One connects customers exactly as instructed. The other knows what its customers want, hires the first company to make needed connections, and bills accordingly. Which company do you suppose will earn more money from an 800-number call? If you haven't guessed, just look at an actual business case. In the early 1990s, MCI rented almost half its circuitry from other companies in order to leave itself free to devote more of its own resources to managing its customers' needs. In other words, if you know what needs to be done, you can frequently get someone else to do much of the internally focused work at close to cost. Competitive advantage lies in knowing as much as possible about customer needs.

Looking toward the future, no sector of the economy can afford to overlook such information opportunities. As the search for competitive advantage in retailing widens, today's mass marketers frequently subdivide their product lines in the hope of finding new niches. But most end up only competing with themselves. In one such case, General Foods now offers over a hundred kinds of Jell-O, I'm told. (I counted several dozen on a recent visit to that part of a local supermarket.) But a consumer's move from one flavor to another makes the move to another brand easier—at least according to business guru Regis McKenna, who contends that brand loyalty must give way to what he calls "relationship marketing." My local supermarket now strives to imitate the special treatment airlines give their frequent fliers. As a result, each of its "Preferred Customers" engages in a personalized transaction at checkout time—essentially exchanging name and address for special coupons (many of them tailored to the purchases made).

Clearly, present-day technology offers enhanced opportunities for entering new businesses via existing customer relationships. My own experience with an early adaptor provides an example. I first encountered United Services Automobile Association, or USAA, in the spring of 1954. A soon-to-graduate college senior, I expected to work for a few months in an out-of-town job before starting two years of obligatory service as a second lieutenant in the U.S. Army Signal Corps. So, in need of car insurance for my newly acquired 1947 Fraser, I contacted USAA. Formed between the world wars by a small group of army officers to provide their comrades with automobile insurance, USAA restricted its clientele to present, former, and (in my case) future commissioned officers in the U.S. armed services. In business for over twenty years by the time I became a member, they had expanded their original charter slightly to include a few other items, such as insurance for boats and trailers.

Since the mid-1950s, however, I have seen an enormous expansion in USAA's offerings. Today, USAA not only offers a full line of insurance—property, life, and casualty—it also offers banking services ranging from credit cards to home improvement loans, a travel agency, investment vehicles such as real-estate partnerships, and even residences for retired members.

Every scrap of information seems put to novel uses. Instead of sending checks to cover the cost of lost or stolen articles, USAA replaces many items with new ones bought at wholesale prices. But they don't stop there. They offer these same items via catalog as part of a buying service. Knowing what its members tend to lose gives USAA a unique sample of their general preferences.

USAA's success hasn't gone unnoticed. A Harvard University study highlighted its phenomenal growth in sales and productive use of resources. Insurance executives I have talked to consistently rate it among the best-run companies in their industry. Among technology providers, USAA ranks high as a sophisticated customer for new ideas. For instance, while it once had clerks assembling stapled copies of insurance records with military precision, it moved to image-based and other electronic forms of document handling well before most of its competitors did. Most recently, USAA embarked on a project to equip its 286-acre main campus for wireless telephony. That way, none of its managers will need to stray from their phones, no matter where their work takes them. Together, these examples of USAA's ability to manage its customer interfaces mark it as an enormously powerful competitor in its chosen fields of operation.

As this example illustrates, a proactive approach to information work supports the establishment of vendor–customer relationships. And these relationships, in turn, support entry into multiple businesses. Taken together, we see a steady progression of impact on business from modern information technology. And each step along the way intensifies the level of global competition.

The genie of international competition has escaped from its bottle and shows no sign of returning. Microchips and worldwide networks have smashed most of the barriers to information flow that authority sought to erect. One need only think back to a besieged Boris Yeltsin in Moscow, giving interviews on a cellular phone from a building ringed by tanks. Even a regime employing millions in its secret police couldn't manage to control its telephone system closely enough to stop him.

But as the so-called Communist countries lose their power to compete with the West in the military arena, economic competition of all kinds escalates. On the one hand, Western enterprises vie with one another for access to the East's markets and national resources. On the other, the People's Republic of China now has a well-educated workforce numbering in the hundreds of millions. Freed from the shackles of a discredited economic system, their entry into a free-market economy may well exacerbate the world's already intensive hunt for new jobs if local demand fails to keep pace with output. While some pundits see globalization as a surefire path to prosperity, competitive risks abound.

Whether as attacker or defender, each enterprise engaged in these arenas will find itself pressed to improve its efficiency vis-à-vis its competitors. But just as technology heightens the need for productivity improvements, it also provides the means for such improvement via productivity gains. While an economy based upon more productive components surely bodes well for our long-term future—as explored in later chapters—it produces turmoil in the near term. The next chapter discusses what happens when employers find that they need fewer people to perform a given amount of work.

CHAPTER 4

REEMPLOYMENT

"Boldly reading the Lord's Prayer backwards and performing many other potent spells for compelling the Evil One." With these words, Charles Dickens in *A Tale of Two Cities* described France's elite—swept along by a revolution they had called forth but could no longer control. Until recently, most of us could afford to invoke the advent of the Information Revolution in whatever terms we liked. Lots of rhetoric, but little impact.

But with the growing merger of computing and communications, a small but rapidly growing number of managers now find themselves faced with more productivity than they had pleaded for. All of a sudden, it seems, work that once clogged an office can move smoothly from its original source to its ultimate destination without intermediate stops. So the hapless occupants of the offices in between find themselves no longer needed. Since the tasks they once performed take no effort at all—and no people—they must seek employment elsewhere.

Competition, humankind's perennial search for advantage, has always been with us. And technology has helped to set the ground rules for work. The lack of an efficient harness drove the Roman Empire to maximize profit by depending on human labor—generally gained through subjugating and enslaving neighboring peoples. Until someone finally devised a collar that allowed horses to pull loads without choking themselves, a given amount of grain produced more work when fed to people than to animals.

Fortunately, today's labor force enjoys a more humane environment in all but a few corners of the world. Indeed, individual initiative has replaced coercion as the key motivator in today's most successful enterprises. As the cases cited in the previous chapter illustrated, top-down-driven hierarchies can't compete effectively against properly empowered teams. The world has become too fast moving and complicated to allow dictation of every move from higher up on an organizational ladder.

But despite humane concerns, no competitive enterprise can afford to employ more workers than it needs. The best figures I can lay my hands on indicate that well-run corporations engaged in high-tech businesses need to grow their revenues by at least 10 (and usually more like 15) percent per year just to maintain the number of employees at constant levels. In order to play in this game, established companies and start-ups alike must continually struggle to enhance employee productivity in order to control costs.

But productivity cuts two ways. As steady progress in manufacturing efficiency and technology advances combine to cut the unit price of finished goods, many businesses find themselves shipping more "stuff" but getting fewer dollars in return.

For example, the size of newly made transistors shrinks enough every month to allow designers to space them about 2 percent closer together. Not much of a change in any one month—but over the course of a decade, that small but steady decrease allows chip makers to increase the number of transistors they deposit on a given-sized sliver of silicon *one hundredfold*. Since integrated-circuit manufacturing costs depend almost entirely on surface area, technology users can look forward to one hundred times more circuitry ten years from now with no significant price changes.

In addition, since smaller-sized transistors generally work faster than the larger ones they replace, each dollar's worth of circuitry will bring along a speed advantage as well. As a result, the same-size chip promises a thousandfold increase in information-processing power ten years from now. But don't forget software. Over the past twenty years, progress in software has meant even greater performance increases than has transistor speed. And this ten-year extrapolation makes no assumption about further scientific breakthroughs. It depends on nothing more than the creative application of our present-day base of scientific and technical knowledge. The point seems clear. Merely holding on to existing business translates into progressively fewer jobs.

Technology and Productivity

Made comfortable by the fruits of our industrial cornucopia, most of us have given little thought to the problem of finding new jobs for railroad firemen, telegraph operators, and the host of other occupations made obsolete by advances in technology. But recent events have brought similar circumstances closer to white-collar employees. Hardly a day goes by without a news report of major layoffs, as one company after another sheds excess staff. Most employers can carry on with fewer office workers, it seems. And productivity climbs. Indeed, 1992 alone produced a greater gain in nonmanufacturing productivity than the three previous decades combined. Good news for stockholders—provided their own jobs remain viable.

Leaving aside for the moment the social implications of mass disemployment, even comfortably employed citizens can't afford to ignore this issue. If growth in productivity continues to undermine future employment opportunities, what should people teach their children? Not an easy question in an age of revolution. With so much changing, can any concerned parent assume that success in life will depend only on admission to the right college?

In the latter half of the twentieth century, a college degree generally offered its owner the prospect of a comfortable work life. Workdays spent in a comfortable office with like-minded people, performing not-too-arduous tasks for adequate pay. If not, one could always switch to something better. But our children will live and work in the twenty-first century. And change aplenty seems in store for them.

Many people speak as if technology's ability to drive change is now exhausted, with no surprises left in store. I dispute that. Instead, I expect a steady stream of technological advances to act as a powerful engine for continued change in the employment picture. As we have seen, the increasing merger of computing and telecommunications has made the work of entire departments unnecessary. And far more downsizing remains possible. After all, the majority of computers in U.S. offices still lack access to networks, and all other industrialized countries lie further behind.

With PCs, faxes, and similar machinery pervasive in most office environments, I can understand the skepticism that greets warnings of further changes to come. Haven't we already automated our offices? In most cases, however, PCs still act as glorified typewriters, helping their users recast text and data. Among other features, tomorrow's workplaces will require machines that can converse fluently with their human users, recognize facial expressions and scribbled text as easily as typed commands, sift through huge stacks of data, and operate seamlessly within a networked world.

Today millions of well-dressed, desk-bound humans still earn a living by "running errands" between machines—getting a fax, marking it up, typing a draft, making copies, filing some, routing others—just to keep the information moving. Much of this work amounts to little more than running errands between one computerized system and another. Hardly a stable situation, especially as direct communication links between computers become more widespread.

Looking Ahead

However powerful and sophisticated machines become, I don't see them as direct replacements of humans—at least not by themselves. Instead, I do see machines enabling a single human to perform tasks that now occupy dozens, and even hundreds, of people—a situation much like that with agriculture during the nineteenth century. In today's industrialized countries, a single farmer now produces at least as much food as thirty farmers did two hundred years ago. In other words, twenty-nine out of every thirty of their descendants moved on to find employment elsewhere.

But a further look into history shows us that such a massive rise in productivity doesn't always produce comparable levels of disemployment. Indeed, a similar rise in manufacturing productivity—as opposed to agriculture—produced a far smaller change in the industrialized West's employment profile. During the twentieth century, manufacturing productivity's steady advance at an average annual rate of 4 percent has resulted in the fiftyfold increase in output per factory worker I mentioned earlier. But manufacturing employment remained relatively stable through that period. While manufacturing occupied some 30 percent of America's workers at its peak in the 1890s, something like 20 percent of a much larger workforce engaged in that same occupation in the 1990s.

On the average, therefore, more than nine out of ten children could follow in their fathers' (and later their mothers') footsteps in each of the last four generations of so-called blue-collar families. Hardly standing still, but nothing like the mass migration from villages to cities that accompanied the advent of mechanical reapers and chemical fertilizers.

So a similar level of technologically driven change produced different results in these two revolutions. In the case of agriculture, mechanization made most of the labor required to grow a given amount of food unnecessary—thereby disemploying massive numbers of people. In manufacturing, the labor force grew from virtually nothing in 1800 to its turn-of-the-century peak, and declined only modestly for the next hundred years. The difference, of course, arises from the amount of output produced.

While individual humans can vary in their food needs to some extent—mainly according to how much of their grain they consume via animal products—biology limits caloric intake. But no similar mechanism caps the human appetite for manufactured goods. From bread machines to aircraft carriers, the number and variety of our civilization's artifacts has, as yet, encountered few limits. Instead, it seems that our seemingly boundless ability to produce goods may challenge our planet's ability to help recycle our trash before most of us finally reach parity with the proverbial Jones family.

Although the long-term fate of manufacturing employment deserves concerned attention, we should remember that some four out of five jobs in industrialized countries lie outside the manufacturing sector. And in less than a decade, people who know how to use information technology to their advantage should have at least one thousand times more power at their individual disposal than exists today. Can anyone believe that such an advance will leave the nonmanufacturing segment of our employment profile unaffected?

Even if one assumes that productivity gains in office work will be shared evenly throughout that workforce, will demand keep pace? Will so-called paperwork track our experience with food, manufactured goods, or something else entirely? If recent experience serves as a guide, future growth in internally focused office jobs seems far from certain. Unlike food and manufactured goods, paperwork confers only indirect benefits on its customers.

As I see it, most of the internally focused tasks that constitute present-day paperwork act like the cogs in a gear wheel. As they expend energy on their side of an interface, they engage their neighbors in the same motions—meetings, memos, and lengthy conversations. Fine (perhaps) for other bureaucrats who need to keep themselves occupied, but what about the folks on the front line? Every minute a customer representative spends meeting with people in her own company reduces the amount of time available for customer contact. Meetings remain an expensive way of transferring information—especially when they tie up people who might otherwise add value.

In the past, office work seemed to take on a life of its own. Consider the amusing but accurate Parkinson's law: "Work expands to fill the time available for it." Or, in our terms, "Productivity declines to keep everyone fully occupied." No more. Corporate boards, acting in response to cries from irate shareholders, have begun to tip the balance toward management that cares more about owners and customers. In the same month that George Bush's one-term presidency came to an end, the CEOs of IBM, American Express, Westinghouse, Wang, and Sunbeam-Oster lost their jobs as well. Clearly, management at all levels can expect more pressure for enhanced performance.

As global communication erases heretofore safe markets, moreover, competition abounds. Virtually every conceivable product and service comes via 800 numbers these days. Under such circumstances, few enterprises can afford to take their customers for granted. Can any afford to employ excess staff? Small wonder that *downsizing* has become a commonplace feature of corporate life in the 1990s. And this downsizing, of course, creates a continuing need for reemployment opportunities.

I recently saw the prototype of a system designed to help sales engineers lay out and sell customized telecommunications services. Today, when someone calls a bank, airline, or mutual fund, they usually get a recording leading the caller toward the desired department, flight information, or current share price. Naturally, the companies involved couldn't buy these systems off the shelf. Each one demanded extensive customization, starting with the initial sales call.

In the past, a sales team—backed by technical support—would meet with a group of their customer's people in order to lay out the desired scenario, so that the vendor could prepare a technical proposal. Weeks would go by. Highly skilled people would turn requirements into technical specifications and outline a timetable for implementation. If all went well, the sales team would then present this proposal to the customer, identify any discrepancies, and repeat the process whatever number of times it took to obtain closure.

What about the new system? A single salesperson can now configure the customer's scenario on the screen of a notebook computer—selecting from a menu, typing in announcements, and accessing databases, right from the keyboard—in the customer's presence. Once completed, this newly created prototype system actually allows customers to test that scenario via a telephone "keypad" that appears on the screen. In that way, they can make sure it matches what they want *their* customers to experience when the actual system goes "on-line."

In actual demonstrations, I'm told, the entire process (including system design) takes less than an hour. Moreover, a salesperson can now take that notebook back to the office and download a complete set of machine-executable instructions that network engineers can transfer to the telecommunications and computing equipment involved—all with no further rewriting of software. In some cases, the salesperson can actually hook the notebook to a telephone line in the customer's office and, after typing in the proper security codes, implement the entire service then and there. The salespeople and the customers can do the entire job without the help of network engineers.

As such systems enter the marketplace, the lesson becomes clear. No job, however sophisticated, can guarantee immunity from technologically driven disemployment. In this case, highly trained telecommunications planners will move to other work, their present jobs short-circuited by a newly established direct link between customer and network control equipment.

Placed in a confined context, direct customer-controlled network management seems like a job destroyer. On the other hand, this dramatic reduction in complexity can make such services more widely available. Why can't individuals get the kind of custom-designed communications arrangements that American Express or Continental Airlines now employ?

Naturally, the price of customization will need to drop way down, but providers should make it up in volume—much like selling water via plumbing instead of by the bucketful. In the same way, while the labor content of each service unit will drop far below present-day levels, vastly increased use should provide more than ample employment for the telecommunication planners mentioned above—at far greater levels of productivity.

Technology's growing power to simplify communications pathways has only begun to make its presence felt in today's job market, and expected advances in technology suggest far greater changes ahead.

But we needn't despair. After all, enough food and other resources exist to feed and clothe all the world's people provided we deploy our assets wisely. From such a perspective, the labor force no longer needed for other tasks becomes an asset that can address as yet unmet needs. In the United States alone, shortfalls in education, housing, transportation, medical care, environmental renewal, and a host of human services might engage a huge labor force, if only one could match skills and resources to needs.

Redeployment Opportunities

For the large majority of today's white-collar workers, reemployment seems highly likely rather than absolutely certain. In this environment, the private sector will likely continue to act as the main source of new jobs. And much of this pool of new opportunities will depend upon new companies forming around new product and service opportunities.

Even companies that merely seek to produce the same old stuff will need to beef up their customer interfaces in order to remain competitive—thereby offsetting at least a portion of job loss elsewhere in their operations.

In all this, of course, fortune will favor those organizations and individuals flexible enough to adapt to changing circumstances. And not only within the private sector. Just look at the costly human infrastructure assembled over the past two generations for the sole purpose of waging the cold war. While some parts of this military-industrial complex have managed to find new employment for their resources—ranging from rides in supersonic fighter-bombers to privacy-preserving telephone encryption—other activities seem too rigidly specialized. But even here, adaptation seems possible. Especially if those consumed see impending change as opportunity rather than threat.

With multibillion-dollar budgets and a workforce numbered in the tens of thousands, weapons-related R&D represents no more than a fraction of an economy measured in trillions of dollars. But, as Senator Everett Dirksen once said, "A billion here, a billion there, and pretty soon you're talking about real money." The redirection of the cold war's R&D

infrastructure deserves attention in its own right as well as being an instructive example of the opportunity side of change.

In the long run, a nation that can afford to devote billions to a space program providing no hint of economic return still has ample room to get its act together. But that doesn't make the process easy, especially for people caught on the wrong side of whatever shifts ensue. What about a physicist with twenty-plus years of experience—and career investment—in a Department of Energy nuclear weapons plant? Would she welcome the prospect of moving to the science department of a small-town high school, provided she could maintain her present salary? While I can't presume to speak for others, I assert that society has a right to examine the question.

Dare we consider closing the DOE's nuclear weapons research laboratories? While I don't necessarily advocate such a step, the thirty thousand dedicated and highly trained people now employed by these heretofore vital institutions now spend something like $100,000 apiece every year—over and above their salaries, pensions, and other personal benefits. Diverted instead to moving and retraining expenses, these funds might enable individual moves into a public education system starved for scientific and engineering talent. At the same time, at least a portion of these sophisticated laboratories will surely find new missions. Created in another era for another purpose, many of them have assets worth maintaining in the Information Age—such as the search for new sources of energy.

Over the years, the scientists and engineers at places like Livermore, Los Alamos, and Oak Ridge have made demanding use of the world's most powerful supercomputers. Should they now direct some of their experience toward creating better supercomputers? Do we even need such supercomputers in an age of PCs? Unless a concerned public and its elected representatives wrestle with the basic issues underlying such questions, present-day momentum may well carry the day.

Practically speaking, economic and political considerations combine to block any meaningful dismantling of such facilities in the foreseeable future. Despite pressure to keep the federal budget deficit within bounds by cutting such expenditures, no representative or senator could afford to stand idly by when payrolls of this size become threatened. Instead, everyone involved searches for plausible needs around which to frame a new mission for such institutions. But whose "needs"? Pressing national concerns that suit available resources, or a massive example of another Peter Principle?

My own experience with government–industry research makes me skeptical about success on such a massive scale. Just look at the money Japan's Ministry of Trade and Industry poured into its heavily touted Fifth Generation Computing project. Today, little remains of this overly ambitious attempt to leapfrog the U.S. computing industry. If government-sponsored initiatives could assure industrial success, the European Economic Community would be sitting on top of the world today. Instead, Europe finds itself without a single first-class computer company on the entire continent.

I do not believe that anyone gains by merely deferring reemployment issues such as what to do with the weapons labs. Once our elected officials recognize the issues with sufficient clarity to reject facile suggestions, they might find themselves with more options than they realize. For instance, at today's prices, I estimate that purchasing individual videoconferencing equipment for every scientist in a typical DOE weapons laboratory would cost less than one-quarter of that institution's annual equipment budget—even less as prices drop over time. With such equipment, individual scientists could share their scientific knowledge with students around the country.

Part-time or full-time teaching, one-on-one tutoring, or formal classroom instruction—the same technology that makes some jobs unnecessary—can also create flexible employment opportunities in this instance. Without leaving their established community and professional roots, DOE scientists can find career alternatives to today's dogged struggle against creeping irrelevance in the post–cold war era.

Easy-to-use videoconferencing needn't remain in the hands of a highly educated elite. Once equipment and connection costs shrink to sufficiently low levels, I see a role for large numbers of learning partners—older students, retirees, home-bound individuals—people with limited formal training who can give academically disadvantaged students the kind of attention now available to children who enjoy more affluent environments. Spending time with each child on an individual basis, such learning partners would certainly repay the investment involved in salaries and equipment.

In considering the negative consequences of technology-driven disemployment, we should not overlook technology's potential for yielding upside gains as well. Fortunately, entrepreneurial ingenuity is still finding new ways of helping people improve their lives. Would you want to dispense with fax machines or cellular phones? Few people saw a need for such items only a few years ago. I think that many more such opportunities lie ahead for technically astute entrepreneurs.

Ingenuity

While many individual problems will no doubt yield to case-by-case fixes, the health of our overall employment picture depends upon a robust private sector, one that continues to create a steady stream of new products and services. But not everyone agrees. Indeed, some argue that little more lies beyond our headlights. The so-called new, they maintain, hardly justifies the effort required to bring it to market. Do we really need talking greeting cards and electric pencil sharpeners?

Back in the days of coal-powered ocean liners and belt-driven dental drills, Albert Einstein described the world as already "too complicated." The thought that anyone would expend *Erfindungskraft* (loosely translated as the "power to discover") on new products struck him as a waste of creativity. While hindsight has demonstrated some shortsightedness in this great thinker's predictive power in this instance, we can't be sure that one day we won't be on a plateau of total satisfaction. But who can say what can't be invented until then? Surely, satisfying these latent needs will provide new job opportunities.

While we can't predict the future, we needn't leave our futures, or those of our children, to blind chance. Much information exists to help us identify likely outcomes of current trends—such as a steady escalation in computing power, for instance. The principal constants of human desire should endure as well—economic security, personal convenience, interpersonal contacts, privacy, and a sense of control over one's destiny. In the Harmony Era, I believe that the match of needs and capabilities will drive our future.

In my early reflections on the transition from the Quality to the Harmony Era, I saw a resemblance to humankind's earlier move from hunting to agriculture. The hunting and quality regimes depend upon the opportunistic exercise of ingenuity, while the Agriculture and

Harmony Eras engage the ecosystem in its entirety. In the future, ingenuity will surely remain a mainstay of progress, but I see us on a surer path than a random search for new businesses.

Over time, companies with the right combination of skill and resources will surely strike it rich—as Xerox did in the 1960s and Microsoft in the 1980s. Just as a few of those lucky enough to get in on the ground floor of those companies became millionaires, so some of today's graduates will find themselves working in places that enjoy a license to print money. At least temporarily. Nothing lasts forever, of course. For if finding gold mines became a surefire process, they wouldn't be gold mines.

Today, almost every industry appears poised for growth. But while we will surely require more VCRs, pretzels, backhoes, and blood tests, none of these items can guarantee growth in jobs or profits for any one producer. Furthermore, productivity gains may surpass growth in demand, thereby forcing a net reduction in employment for that industry as a whole. Under such circumstances, competition between look-alike products usually squeezes prices to the point where staying in business becomes an exercise in pain endurance.

As we saw earlier, the transition from quantity to quality hinged on making better versions of similar items—automobiles, television sets, or cameras. If the harmony concept proves correct, the most promising areas for business growth won't come from head-on assaults against existing enterprises. Profitable as such actions might be for the winning team, their success would merely shift jobs from one group to another. While head-to-head struggles go on daily in today's highly competitive environment, I see end runs into unaddressed areas as a more likely locus of real growth.

As I've said, ingenuity can bring a "hunting image" to mind: the discovery and exploitation of a heretofore neglected area, such as in the introduction of fat-free frozen yogurt. As ever more offerings enter the market, however, less and less open space remains. And that, of course, reduces the hunters' chances.

From my perspective, I see these narrowing gaps between existing business areas as opportunities for profitable bridge building, as in the Certified Collateral Corporation example we discussed earlier. Its founder built a multimillion-dollar business out of a two-way flow of information between insurance companies and used-car dealers.

On a more global scale, take the energetic entrepreneurs who found billions of dollars' worth of business opportunities along the fault lines marking the yet incomplete merger between computing and telecommunications.

The business created by two Stanford University computer administrators, Sandy Lerner and her husband, Len Bosack, offers a fine example. Working for different departments in the mid-1980s, they couldn't swap electronic mail and software because their respective computer networks lacked a way of communicating with one another. When Bosack devised and built a means of linking the two networks together so they could share software and databases, the couple realized they had the beginnings of a truly valuable product. But none of the established companies they approached showed interest. Fortunately, Lerner and Bosack didn't take no for an answer. They proceeded to quit their jobs and launched Cisco Systems in their garage.

Always short of cash, they took out a second mortgage and stretched their credit cards to the limit. But the positive reception their *router* received from networking colleagues convinced them that they were on the right track. Sales climbed rapidly, but so did expenses. So they found themselves forced to give away stock to pay bills, and by 1990 they had lost control of their company. Both left as a result of a dispute in August of that year and sold their remaining shares for slightly in excess of $200 million dollars.

The folks who bought that stock didn't do badly either. From $27.7 million in 1989, Cisco's sales topped the $1-billion-per-year mark a mere five years later. While established computer companies struggle to break even, Cisco consistently nets eye-popping returns on sales, generally in the 25 to 30 percent range. Small wonder that Wall Street valued the company at over $5 billion less than ten years after its birth as a "mom-and-pop" start-up.

By identifying an information gap and building a bridge across it, Cisco's founders spurred the merger of computing and communications. While computers can exchange data within an office building or campus via local area networks, these LANs have a hard time communicating with one another. In order to get good performance over short distances, LAN designers opted for systems parameters that prevented operation over longer distances. Since electromagnetic energy can't travel faster than the speed of light, round-trip delays loom large when measured in the billionths-of-a-second intervals at which computer systems pace themselves.

Today's *routers*, as they are called, switch data traffic from one LAN to another over transmission lines leased from telecommunication companies. Users on geographically dispersed systems can swap messages with one another so conveniently that *local* area networks are quickly becoming a thing of the past. While the LANs themselves continue to exist, their links to the larger world permit networked computer users to gain access to information on a worldwide basis—a subject we'll explore further in a later chapter.

Summing Up

A move of job opportunities toward business interfaces seems increasingly evident. With the hot breath of competition forcing businesses everywhere to enhance workforce productivity, few jobs remain untouched. Change pervades the computer centers where the "organization man" of a bygone generation once held sway. Some departments no longer exist, their functions made superfluous by information bypass. Disemployed workers generally find new work closer to customer interfaces, either in the same corporation, in other companies, or as individual consultants.

Even those whose jobs appear to remain intact rarely escape the forces of change entirely. In my own case, while many of our organization charts look little different from their 1980s equivalents, a lot has happened to the behavior and attitudes of the people in my company during the interim. In particular, we pay a lot more attention to external matters—especially as they relate to customers. Just a few years earlier, many researchers saw business concerns as distractions from their "real" work. Today, those same folks complain when they can't get their ideas incorporated into products.

Internal issues get less attention in healthy companies these days. Old corporate models no longer apply. Traditional bureaucracies no longer provide a steady supply of job opportunities, returning surplus workers to the job market instead. So as displaced workers seek reemployment in a changing workplace, new styles of work engage more and more attention.

CHAPTER 5

WORKSTYLES

How is the need to look outward reshaping our workplaces? As managers struggle to adapt themselves and their organizations to a changing business picture, they find themselves caught in a dilemma. On the one hand, traditional hierarchies allow a small number of people to coordinate large-scale, complex tasks. On the other, these same hierarchies seem poorly positioned for entry into new opportunity areas. So, while the vertical workstyle offers some advantages, the need for new approaches seems evident. Organizations designed and motivated to address internal issues must now shift their focus to their borders. Reemployment, therefore, becomes more than a matter of filling the same old jobs with new employees.

Given a choice between beating one's brains out in head-to-head competition over existing terrain and staking a claim to the next Cisco Systems, most would opt for the latter. But a wholesale move to garage-style operations won't solve all our reemployment problems. For every Cisco Systems or Microsoft, uncounted numbers of start-ups have stumbled and failed—just as some established businesses have found paths to brand-new multibillion-dollar opportunities. Simple prudence suggests that we examine hierarchical workstyles with an eye toward correcting their faults rather than just pinning all our hopes on more radical alternatives.

In the past, keeping each organizational fragment insulated from the others made sense because horizontal communication weakened overall coherence. In a system designed to deal with complexity by controlling it, any tendency to look outward made control more difficult. So the reward system encouraged internally oriented behaviors. But as we have seen, that "simplification" ultimately laid the groundwork for the system's near demise. In our earlier mainframe-versus-workstation example, we saw how a computer company's compartmentalized information flow thwarted a needed new-business initiative. The hierarchy in that case promoted parochial interests at the expense of the larger corporate good. And at considerable cost.

Faced with abundant examples of similar behavior, we shouldn't wonder that some organizational theorists question the continued viability of hierarchical management structures altogether. Peter Drucker, for example, depicts an organizational ideal in terms of a symphony orchestra—dozens of musicians, all engaged in their specialties under the baton of the single conductor. But such coordination suffices only as long as all concerned follow the composition on a note-by-note basis. Little room for managing the unexpected complexities that accompany moves into new territory.

In today's business world, new forays into uncharted areas frequently require such a broad range of skills that even the largest corporations can't deploy all the needed expertise by themselves. At AT&T, for instance, we have formed partnerships with companies ranging from Apple to Zenith in order to address various new opportunities. In each case, the folks involved had to break new ground in a host of specialized areas, while maintaining coherence across the entire project. The mind-boggling number of issues involved required constant coordination among team leaders. So that team of team leaders required a leader of its own, and so forth. There seems no evident way of avoiding the need for some sort of structure. But what form should that structure take?

In Search of Paradigms

Chairing a 1994 symposium on the future economy, University of California professor Paul M. Romer asked the following question: "What image will replace the factory: office park; film studio; research laboratory; software house; university; or none of the above?" Good question. As one of Professor Romer's panelists, I appreciated the fact that he faxed it to us ahead of time.

At first, it seemed like an easy assignment—pick one or two from the list and argue in favor of my choice. Not like school, where only the right answer got me a passing grade. In the end, however, I found myself unable to make a selection. All of the images seemed valid in some respect. With information as the key component in value creation—that is, doing something for which someone else will pay money—large numbers of purposefully employed humans in the economy of the future will surely acquire, process, create, shape, share, and deliver information. But which model suits this task best?

Office park. Today, the majority of people engaged in information tasks seem to work in offices. As one looks to the future, it seems reasonable to surmise that such activities will continue to center upon an officelike environment.

Film studio. On the other hand, everyone is drowning in piles of paper already. Who needs more reports? In order to make tomorrow's flood of information accessible (and thereby economically attractive) to its intended consumers, tomorrow's providers will surely devote themselves to creating more tractable formats. They'll use visually attractive graphics accompanied by well-prepared scripts, much like what we find in today's film and television studios. So tomorrow's information providers might arguably work in the same sort of physical setup as a film company does with its soundstages and other production facilities.

Research laboratory. Since superior knowledge will presumably confer decisive competitive advantages upon its possessors, we might well expect future "capitalist" enterprises to concentrate on building knowledge bases—either for internal use or for sale to others. In that sense, we see an increasing number of "knowledge factories"—institutions designed to produce knowledge. And, in today's world, research laboratories come closest to filling that role.

Software house. Looking ahead, it's hard for me to imagine a product or service without a significant software component. Even where the final consumer may not see software in the offering in question, virtually every business process involves programmed machinery of some kind. With machines performing most of what people used to call "work," a great portion of human labor will inevitably shift toward giving instructions to our electronic helpers.

University. While both the research lab and the university "manufacture" knowledge, the latter embraces a broader range of disciplines and includes formal learning among its primary objectives. Since value creation in the Harmony Era will involve the mastery of wide-ranging interactions, breadth of knowledge will play a key role. Ongoing learning will help to preserve and enhance any future "factory's" primary asset, its expertise. While formal classroom instruction should continue to play some role in knowledge dissemination, I see the doing aspects of university life—notably individual and group scholarship—as closer to this future workstyle model.

None of the above. Finding attractive elements in each of Professor Romer's candidates, I also noted that none of them appears to enjoy robust growth in today's period of transition. Moreover, each of these items dealt more with a piece of the puzzle than with an integrated whole. And, as I noted earlier, value creation in the Harmony Era will emphasize integration and customization. In that spirit, a bit of head scratching led me to propose my own addition to Professor Romer's list:

Architectural firm. Ideally, an architect should provide a customized solution to an individual's needs, one that is integrated within its environment and in conformance with the requirements of the systems with which it comes in contact. Consider the blend of customization and integration that you would expect from an architect hired to build your family's dream home.

> *Cost:* Works within your budget and negotiates as appropriate—stretching to add desirable features, weighing, trading, and balancing each expense component.

> *Customization:* Engages you and your family in dialogue to understand your needs, preferences, and general lifestyle. Low-maintenance lawn or elaborate landscaping? Room for big noisy parties? Quiet nooks for individual reading? Home offices? Pets? Hobbies?

> *Environment:* Creates a design in harmony with its surroundings. Natural light. Solar heat for colder seasons and locations, together with shade and cooling breezes when the thermometer rises. Efficient use of energy and other resources.

Integration: Conforms to mandated standards (such as building codes) and links smoothly with services and surroundings. Heat, water, electricity, waste disposal, access, neighboring structures.

Esthetics: Pleases the senses with an attractive total package. Views, both of the house and from it. Materials and styles that suit its surroundings and adjacent dwellings. A living space that comforts and lifts the spirits of its occupants.

Sounds like a tall order, even as a metaphor. How can anyone afford to put this kind of effort into mass production items? Admittedly, we can't afford much customization in small items like paper clips, but I expect to see everyday TV sets with many of the customization elements we now find in upscale homes.

In this latter case, I speculate that a personal (or electronic) visit to your favorite appliance store to buy a new TV set will involve multiple opportunities for choices. Instead of just specifying screen size and choosing among available models, you might design a remote control to suit your own control vocabulary. Bring along your old controller. If you like it as is, fine. The new set will respond to its commands. If not, add, delete, and relabel: buttons, features, and whatever verbal command functions you would like to employ. In addition, your new controller will operate your existing VCR, stereo system, communications gear, and the like, while the new TV set will play into—and play with—the family of appliances you already own.

In those cases in which customization and systems integration merely involve software changes, the needed code can either be supplied by the retailer or downloaded from the manufacturer's inventory of optional software modules. Similarly, if the package requires any extra hardware to produce asked-for features, the dealer would generally install additional plug-ins into the unit. In some cases, however, the purchaser might ask for a custom creation from the factory—a unique item flowing down a production line so flexible as to accommodate unit lot sizes, much like the made-to-measure jeans that Levi Strauss began offering in the mid-1990s.

In this example, much of the value of the final product arises from the interactions involved in customization. In the future, I expect vendors to compete with one another on the basis of their ability to customize all across the economic scale, on much the same terms as now exist for larger (and therefore individually designed) projects—such as the dream house I described above. So boundaries between functions in the Harmony Era will keep blurring. Fewer and fewer institutions can afford to say "My job stops here" and maintain a healthy business.

At the same time, our harmony concept demands coherence along with breadth. To me, the workstyle we now associate with architecture provides a tangible foundation for the disparate themes we can expect to find in tomorrow's workplaces. As we speculate on the future, therefore, I see architects as quintessential harmony-age workers. This profession, more than any other, brings coherence to mind. And, since coherence provides a key theme for the Harmony Era, let's pursue this model a bit further.

Architects

As we explore corporate structures, we shouldn't overlook the coexistence of other organizational loyalties in these very same institutions. In particular, architects—like physicians, purchasing agents, engineers, film directors, and teachers—belong to professional societies. Together with less formal ties, professional affiliations have long served to influence behavior and to permit the sharing of information among their members, thereby competing with, and supplementing, management's traditional role.

While I use the architect as a concrete example to which general readers might relate, more technically minded readers will recognize "architecture" as a buzzword in systems integration—the practice of getting machinery and software of all kinds to work together.

Many firms specializing in systems integration resemble conventional architectural firms in their organizational makeup. Just as some architects work as individual entrepreneurs, either alone or with partners, we find numerous examples of similarly organized consultants who lay out information-processing and telecommunications systems for their clients.

While such consulting practices normally start small, some, such as Andersen Consulting, have grown to remarkably large size. Begun as a service by the accounting firm of Arthur Andersen, this systems-integration practice grew to a $1-billion-per-year enterprise by the time it split off as a separate business in 1988. Moreover, partnership work-styles its founders had established proved no obstacle to further growth. Over the next five years, Andersen Consulting tripled its annual revenues to some $3 billion—making millionaires out of many of its 858 partners, according to *Fortune* magazine.

While Andersen Consulting engages in "manufacture" in the sense of creating applications software for its customers, we can't expect consultants to build automobiles, provide overnight mail-delivery service, or rent videos. Their relatively loose organizational structure doesn't lend itself to the tight coordination needed to run Nissan Motors, Federal Express, or Blockbuster Video. At the same time, however, conventional businesses will need increasing amounts of the services currently offered by such self-employed professionals.

Generally speaking, we expect consultants to bring a body of knowledge to their tasks and to take responsibility for keeping that knowledge current. The short shelf life of knowledge today mandates a continuous commitment to learning. While compartmentalization called for the kind of expert who knew more and more about less and less, the Harmony Era demands an ability to deal with wide-ranging interrelationships—as with the "simple" task of designing a truly satisfactory single-family home.

On a larger scale, systems integration for a modern corporation involves the needs of associates spread around the globe. Here, a new release of networking software has great impact. Unexpected combinations abound. Consider the impact of a special word processor bought to accommodate a global corporation's Cairo office. What happens when Arabic text encounters electronic forms that expect right-to-left character strings?

In such an environment, the need to deal with unforeseeable circumstances will favor an ability to learn and explore as well as the broader perspective that comes from standing back a bit—functions traditionally provided by architectural consultants. As time goes on, however, successful institutions must incorporate similar capabilities within their own workforces or cede the bulk of value creation to others. The "architectural firm" itself, in other words, must add value to survive.

While tomorrow's workstyles must avoid the internal focus of old-fashioned top-down control, the task of creating a coherent whole out of individual contributions remains paramount. How, then, to reconcile these two requirements? As I see it, future hierarchies must revamp their reward systems to de-emphasize corporate ladders—rather than attempt to abandon those ladders entirely—so as to encourage outward-looking behaviors in those who strive to climb.

Climbing the Ladder

From early childhood on, our children are taught to equate progress with promotion. Playtime, nursery, prekindergarten, and kindergarten—a series of steps tracks the child's growth and prepares each student for the "real" ladders that lie ahead.

A playground conversation I overheard while in graduate school (twenty-first grade, by my calculation) captured some of this early training. "Are you in first grade?" an evidently awestruck preschooler asked a taller companion. "Higher than that!" the other replied, puffing up his seven-year-old chest with pride. Given the experience gap between them, that second-grader wisely refrained from giving a detailed explanation of the steps that lay beyond the younger child's horizon—let alone the means of ascent.

In today's world, unfortunately, the "means of ascent" frequently translates into counterproductive behavior. How? Corporate reward systems often sanction inefficiency, because so much of "rank" depends upon the number of people in some pyramid.

Let's look at the accounts-payable departments of two similarly sized corporations. One employs six people operating a networked desktop computer system linked directly to vendors, while the other employs five times that many people, including four supervisors, to do the same job. In addition, the latter operation requires an expensive mainframe. Which department's manager, do you suppose, would command a higher level of reward? In most cases, the "greater responsibility" of running the larger-sized organization would give that department's head a larger salary as well as improved chances for promotion to yet-higher positions.

Improving the latter organization's efficiency, therefore, may well bring the risk of demotion to its leaders. Small wonder that most middle managers fear the loss of status that frequently accompanies new workstyles. Rank, as they say, has its rewards.

Herman Wouk captures this rank consciousness in his *War and*

Remembrance, a sweeping historical novel of the Second World War seen through the eyes of a U.S. Navy captain. If fictional characters ever gathered at some mythical cocktail party, Wouk's hero could have dropped names with the most traveled of his compatriots. Hitler, Churchill, Stalin, and Chiang Kai-shek—Captain Victor Henry meets them all, thanks to a series of missions for Franklin Roosevelt. But Henry keeps fretting. Will the next assignment bring him closer to the command of a battleship? Even a front-row seat at the greatest drama in modern history couldn't divert his attention from the career ladder he grasped on his first day as a midshipman.

We may smile at such concerns, but striving animates achievers in most organizations. Like the U.S. Navy, therefore, corporations still count heavily upon promotion as a key incentive. But downsizing, and the evident economic advantages that stem from flattened organizations, combine to curtail promotion prospects. So the majority of today's entrants into the job market can't look to rank for reward, at least not in the traditional sense. Instead of mastery over others, future managers might better look to mastery over circumstance—problems solved and skills gained.

At the risk of wishful thinking, I speculate that the power to accomplish a complete task may compensate for the "power" that once came with gold-braided sleeves. In bygone days, every layer in the hierarchy occupied itself with massaging messages as they moved from one specialized group to the next, so work offered relatively limited satisfaction. With never more than a piecemeal part in any task, no wonder that yesterday's organization man looked to organizational status for reward and guided his actions by whatever might most please his boss.

But today businesses need to encourage people to look outward rather than upward. In this environment, work becomes more team based, with coaching—rather than detailed direction—from "above." Consider the Compaq salespeople I mentioned earlier. While each one works from home and remains free to schedule the day without asking some boss's permission, none works in isolation. All relevant information lies within immediate reach. It just takes a few keystrokes.

With all of Compaq's information systems tuned to supporting the company's sales effort, I can imagine a new kind of pyramid. Instead of shipping messages up and down through multiple layers of subordinates and their superiors, this new arrangement shapes and shares information for the benefit of the person at the customer interface. Here, the "ladder" lies sideways. In this way, individual empowerment encourages collaboration with colleagues and service to customers instead of rote response to top-down directions. Hopefully, we can shape tomorrow's work environment to offer a greater supply of similarly empowering opportunities throughout our economy.

While promotion will still remain an important career path, the focus of jobs at every level must shift to accommodate an outward-oriented perspective. Instead of just collecting information, organizations must put it into the hands of the people who need it—such as top-line executives in the workstation example and customer representatives in the Compaq case. While the latter occupation's importance continues to grow, as we've discussed, the former remains important as well. After all, redeploying Compaq's salesforce required an informed top-level decision.

Here, then, we see a bridge between existing and emerging workstyles. As organizations apply multiple skills to a complex task, information must flow in a coherent manner from the specialists who have it to all the folks who need it—top managers as well as frontline troops.

Coherence

In order to justify its existence, an organization must add more value than would an equal number of uncoordinated individuals. An organization, then, succeeds or fails according to its ability to bring the benefits of specialization to bear upon a given task. And that requires communication.

For our purposes, we may think of an organization's information flow in terms of three needs: distributing a complex task among specialized contributors; delivering their combined output to customers while receiving feedback from them in turn; and supporting management's coordination function. As we trace the evolving relationship between form and function, therefore, we need to understand the relative importance of each of these information pathways with various kinds of tasks.

In the mechanical era, as we have seen, specialization generally meant a sequence of separate steps, each one different from its neighbors. And not just in factories. I know of at least one major insurance company that, as late as the early 1980s, required over fifty different approval steps for each new policy. Here, hierarchical partitioning of the tasks involved, as well as minimal contact with customer contacts at the working level, suited top-down planning and coordination just fine.

As we moved to the Quality Era, on the other hand, programmable flexibility supplanted rigid mechanical design. Tasks could change over time, and subtasks might move from one stage in the process to another. Information had to flow horizontally to help allocate functions and coordinate the work. Contact with customers grew, and a team-based workstyle emerged. Action didn't wait for theory.

In today's environment, new workstyles are emerging well before organizational theorists can give them labels. As innovative humans try various approaches and imitate those that appear to suit them best, I speculate that some combination of global coordination and local teamwork will emerge. In this connection, most of the managers I see these days spend the largest portion of their time looking sideways—that is, meeting with peers and customers—rather than dealing with the parochial concerns of their particular specialty.

My own experience at Bell Labs mirrors this trend. Like that of most of its corporate contemporaries, Bell Laboratories' organizational makeup employs a familiar multilevel hierarchy. Moreover, our long tradition of specialization has weighed heavily against radical changes in the status quo. As recently as the mid-1980s, therefore, most researchers worked to enhance our owners' chances for long-term success as we always had—alone or in small groups, and only working with others when a project required outside help—all the while maintaining the organizational style that had served us so well in the past.

As time went on, however, it became clear that this traditional mode of operation had failed to meet our expectations. Bell Labs wasn't providing an "unleashed AT&T" with the dazzling array of new products that pundits had predicted in the heady days following the old Bell System's breakup. What to do? For a while, we contented ourselves with waiting for our colleagues to patch the leaks in *their* part of our corporate boat. When we finally took a dispassionate look at ourselves, however, we found traditional behaviors ill suited to the real needs of our enterprise.

Not unlike departments of other compartmentalized organizations, each research group worked to exploit opportunities as its people saw them, with little attention paid to others. One area of technology might receive attention from multiple groups, while another might receive none at all. The situation called for change. No matter how well the past work-styles had served us, they no longer sufficed. But that didn't mean trashing our organizational structure. Instead, we concentrated on giving each part of the organization a needed, outward-looking focus. Among other things, each research manager had to take on an additional assignment. Instead of just striving to make sure that each scientist and engineer in a given department worked at the highest level of accomplishment, managers took responsibility for fulfilling our company's technology needs in areas assigned to them.

The change didn't take place overnight, but over time, our behavior has changed radically. Today about half of our researchers work full-time in partnership with colleagues from other parts of AT&T. Similar changes have taken place in management as well. While most research managers have kept their titles and the trappings of office, their jobs have undergone ninety-degree turns. Instead of looking up and down, so to speak, they now spend most of their time looking sideways.

For example, each laboratory director now works with one of AT&T's business units, making sure that its needs get attention and that researchers have access to potential customers for their work—not just for the people in that director's organization but for the research operation as a whole. With organizational roles now more clearly defined on the basis of function rather than scientific discipline, management's primary attention has shifted to external interactions.

Recasting first-level management roles has proved the most challenging undertaking. Experienced researchers themselves, managers had worked hard to ensure the best possible research in their departments. But "best" as they defined it: the world's most powerful laser diode; a record-breaking transmission experiment; the "best paper" award at a major professional conference. While certainly not unworthy, this internally generated pursuit of excellence paid insufficient attention to the priorities of potential customers. While colleagues in our Lightwave Business Unit sought more powerful lasers, they might have preferred to trade some of that device's performance for compatibility with their existing fabrication methods.

Since the researchers saw themselves as guardians of traditional excellence, they naturally regarded new criteria as a lowering of standards. But over time, the new ideas took hold. Building a *manufacturable* record-breaking laser presents a far greater challenge than building one that just works long enough to get a paper published, after all.

The new mode of management employs much of the old style and value system. Everyone still strives to be "the best," but under new and more demanding criteria. Attention that once focused largely on internally generated standards of job performance has shifted toward interorganizational teamwork at all levels. Quite a change from the days when management sought to protect researchers from untoward contacts with the "outside." Like our corporate colleagues, we strive to adapt ourselves to a new era, one that combines a small-scale entrepreneurial spirit with the large-scale coordination required by massive global enterprise.

Size

Individual units can change their behavior without changing their structure, but what about the corporation itself? With former economies of scale offering less than certain advantage, does sheer size present a handicap? Will start-ups ultimately defeat themselves by becoming billion-dollar monoliths? Or should growing businesses constantly fragment themselves to keep from getting too big for optimal performance?

With so many Americans employed by large corporations, either directly or indirectly, and with so much of our money invested in them, the long-term success of big business weighs heavily in the future of our economy. Yet past practice falls far short of the mark. In former years, Alfred P. Sloan, president of General Motors, was universally hailed as an organizational genius for building GM's management structure. Yet those who followed in his footsteps piled up staggering losses. In less than a decade, GM's sales dropped by some three million cars per year—enough to fill a two-lane road between Detroit and Osaka, bumper-to-bumper in both directions.

"Didn't listen to their customers," the pundits proclaim whenever a large company gets into trouble. The unlucky executives involved usually deny the charge, but conventional wisdom generally associates an organization's large size with insensitivity to its surroundings. And, as a matter of fact, the laws of physics lend some support to this notion.

When Hannibal crossed the Alps in 218 B.C., this celebrated Carthaginian general led an army recruited from North Africa and Spain, complete with elephants. Proceeding across Gaul, he marched southward over the Alps to attack Rome from the landward side of the Italian peninsula.

While Hannibal won several famous battles, the campaign turned out badly—especially when some of his elephants panicked and charged in the wrong direction. While from a military point of view, Hannibal would have been better off leaving his elephants at home, the image of these huge creatures marching through the snow amidst a horde of evidently freezing soldiers has given us one of ancient history's most enduring images.

This historic encounter with cold weather didn't seem to bother these walking tanks—even though they had come from tropical climates. Why? Because every square centimeter of elephant skin exposed to the elements could count on many times more body mass to replace lost heat than could that of their human counterparts. Sheer bulk, in this case, reduced the influence of the external environment.

Today, the remains of some of yesterday's monolithic organizations still find themselves insulated from their environment for much the same reason. When the amount of time each employee expends on internal issues dominates, many in America's largest corporations still turn a blind eye to the changes occurring around them. Small wonder that an organization man could once look forward to spending the entire interval between graduation and retirement behind a desk. Sheer organizational size, it seemed, provided much the same insulation that Hannibal's elephants enjoyed.

Still, size may not be quite as much of a handicap as this image of a corporate elephant might suggest. "The size doesn't matter," as one physicist described it. "Just the surface area. It's the difference between a tightly packed snowball and a cloud of snowflakes." While I'm not totally comfortable with the metaphor, it makes the point that *shape* (in this case, the distance between individual employees and their customers) rather than *size* (the total number of people in an organization) determines the level of contact with the external world.

As I noted earlier, abundant means of production have driven businesses (and other institutions that depend upon outside constituencies for their survival) to compete on the basis of customization. In that framework, customer requirements become a competitive organization's most critical domain. All other resources exist to support them. As a result, each corporate organization needs to prune its "insides" so as to enhance its surface area. Unlike the organization man who pursued his career safely ensconced in a corporation's inner workings, tomorrow's job seekers would do better to link themselves to the locus of action—just as organizations will perform best when coupled to their environments as efficiently as possible.

In Bell Labs' case, only a shrinking minority of research physicists study "physics" today. While we continue to pursue scientific excellence in fundamental areas, most of our scientists seek breakthroughs in targeted areas—such as the special lasers that have brought lightwave technology to the cable television industry. With billion-dollar contracts riding on who can supply the best technology at the lowest price, today's scientists can hardly afford to behave as they did when we were the only game in town. While our size remains the same, therefore, our customer environment gets a far greater share of our attention.

I see quite the same kind of change throughout the business world. So irrespective of size, therefore, we would expect successful enterprises to shape themselves more like snowflakes than snowballs.

Indeed, a number of formerly monolithic corporations have divided themselves into groups of smaller, largely autonomous business units. But whatever their size, organizations have the freedom to change behavior as well as structure. For example, direct customer contact at the working level translates into less need to check up the line of management and thereby leaves more time for interacting with the organization's "outsides." So one of the most effective business leaders I know interrupts any meeting that goes longer than ten minutes without mentioning "customers."

Today, every successful company must strive to delight its customers. Otherwise they'll go elsewhere. Under such circumstances, vendors work to tailor their offerings as closely as possible to customer desires. Business publications use buzzwords like "relationship marketing" and "lot size of one" to describe such efforts. I have a simpler image in mind: the customized, flexible service offered by an individual entrepreneur—the ultimate in small companies—but with the broad resources and mix of skills expected from a world-class operation.

The more an organization devotes its resources to customer contact, the more monolithic workstyles must make room for more flexible employment relationships. While increasing numbers of workers will need ready access to wide-ranging information resources, optimal information sharing needn't demand one's physical presence in some corporate office complex. One of the researchers in our Murray Hill, New Jersey, Computer Science Research Laboratory works in an office in a corner of his Seattle home much of the time. Another telecommutes from an AT&T facility in Taipei. Tomorrow's interactions may depend far more on shared understanding than on sharing the same office corridor.

On a superficial level, most twenty-first-century enterprises still seem likely to employ large numbers of professionals as "office workers." But the definition of "office" should broaden considerably. Imagine yourself as an employee in one of tomorrow's "architectural firms." What does a conventional office offer you? A typical list of features might include:

Work-in-progress—a half-finished report, letters in need of reply, stacks of reading material, and the like.

Access to people—the formal and impromptu meetings held in and around the area.

Access to facilities—computers, copiers, fax machines, and all the other necessities of modern office life.

A messaging gateway—entry point for mail and phone calls.

The primary store of local information—paper files (and piles), as well as local databases.

I could go on, but a longer list would support the same observation: Each item now lends itself to increasingly effective electronic substitutes. The advent of electronic writing tablets, optical character—and handwriting—recognition, electronic storage of documents, and the like will keep all the "papers" you need as close as your personal (computer) notebook. Moreover, that same notebook will surely act as your phone as well—for both voice and video, of course.

As I see it, the demands of closer connection to customers, partners, and suppliers, as well as to professional colleagues, will surely keep a considerable portion of our future workforce away from "the office" at any given time. While modern communications will reduce the need for travel in many instances, I can't imagine fully satisfactory electronic substitutes for all forms of human interaction. And travel leaves less time for sitting behind a traditional desk.

Since an ever greater portion of business interest seems destined to shift outward to corporate boundaries and beyond, each of tomorrow's organizations will need to make its resources available to geographically displaced members of its workforce as well as to the multiplicity of organizations and individuals with whom it will relate. In such a dispersed environment, a management's ability to create coherence through shared vision and a common set of professional values will rank among its most important competitive assets.

Since this mix of work and lifestyle will demand global communication links among work, home, school, and even recreation, our so-called information highways become an issue of national significance.

CHAPTER 6

NETWORKING

On February 8, 1993, barely three weeks after William Jefferson Clinton took his oath of office as the forty-second President of the United States, the following electronic mail message appeared on my computer screen:

Subject: Send E-mail to the White House

The following is NOT a joke. It is a real service being provided by President Clinton. If you feel you have something to say, send them a letter. I heard they will reply to most messages.

As you know, this is the first time in history that the White House has been connected to the public through electronic mail.

We welcome your comments and suggestions for ways to improve your Public Access E-mail program.

Jack Gill

Electronic Publishing

Public Access E-mail

The White House

Washington, D.C.

——End of Forwarded Message

All replies, however, are through regular mail, so be sure to include your mailing address.

(The White House e-mail address is that text between the <>:
<75300.3115@compuserve.com>)

——End Included Message——

The White House, it seemed, had obtained an e-mail address, thereby offering the general public an electronic pathway to the world's most powerful office.

Why didn't the world's foremost chief executive rate a glitzier address? I wondered.

As the employee of a *government* agency, *President* Clinton subsequently had someone on his staff acquire a computer for the *White House,* tie it to the public network described later in this chapter, and give his incoming e-mail a name. "President@WhiteHouse.gov" seems more descriptive than "75300.3115@compuserve.com," I'll admit, but the latter does have a more populist ring to it.

Communicating by electronic mail involves shipping a *file* stored in one computer to some other computer via a *network.* While most such files consist of a few lines of typed text, files can also include such things as stored images and diagrams, tables of numbers, or even sensory data, such as an electronically recorded electrocardiogram.

Strictly speaking, e-mail refers to file transfer controlled by an electronic mail program. So it represents only one of many methods of intercomputer information exchange. In the case of e-mail, a program in the sending computer stuffs mail messages into the electronic equivalent of envelopes by adding identifying markings, such as the sender's address. These electronic *headers* allow a companion program at the receiving end to collect and sort incoming mail for reading at the recipient's leisure.

E-mail can also act as a building block for more complex arrangements, such as computer bulletin boards. Here, a group of like-minded individuals might send e-mail to an agreed-upon location, one that supplies a copy of all its stored mail to anyone requesting it.

E-mail is quicker and less formal than paper and easier to scan and edit than recorded voice messages. Since it avoids the need for a recipient's presence at the other end of the information-transmission path, many busy people prefer it to playing telephone tag. Small wonder, then, that it has become a valued means of keeping in touch for a worldwide computer-adept community, one which now numbers in the millions.

Today, millions of PC users dial up services via ordinary telephone lines and exchange data by means of their *modems* (modulator/demodulators), devices that convert data pulses into the kind of speechlike tones that telephone lines normally accept. When connected, modem-equipped personal computers offer far more than just a means of opening an electronic mailbox. Access to electronic bulletin boards, browsing through shopping services, libraries, and the like, can all be yours as you explore today's fascinating agglomeration of quasi-public networks.

In keeping with promises made on the campaign trail, the Clinton-Gore administration has placed high priority upon the creation of a so-called Information Superhighway. Given the manifest benefits of America's interstate highway system, the image of an Information Age replica finds ready acceptance from press and public alike. Just optical fibers instead of concrete.

But on closer inspection we can see that the highway metaphor highlights transmission facilities at the expense of other equally important components. After all, end-to-end information movement calls for items that vehicles don't require—such as applications software. So while the "highway" metaphor remains firmly fixed in our minds, further planning has broadened the concept. Today, technology planners speak of a "National Information Infrastructure (NII)": a web of communications networks, computers, databases, and consumer electronics that will put vast amounts of information at users' fingertips.

Expanding the Superhighway Concept

This NII concept aims high. In addition to expanded "communication" and "library" services, government, public-interest, and industry leaders have identified *education, medical care,* and *industrial productivity* as particularly important targets. In other words, we can expect enhanced support for schools, hospitals, and industrial design centers. For instance, NII literature depicts local physicians supported by specialists in distant cities via videoconferencing and medical image transmission. On a broader scale, a September 1993 working paper by the Clinton administration's Office of Telecommunications Policy depicts the promise of the NII by asking readers to visualize the system's benefits in the following terms:

> Imagine you had a device that combined a telephone, a TV, a camcorder, and a personal computer. No matter where you went or what time it was, your child could see you and talk to you, you could watch a replay of your team's last game, you could browse through the latest additions to the library, or you could find the best prices in town on groceries, furniture, clothes—whatever you needed.

Imagine further the dramatic changes in your life if:

The best schools, teachers, and courses were available to all students, without regard to geography, distances, resources, or disability;

The vast resources of art, literature, and science were available everywhere, not just in large institutions or big-city libraries and museums;

Services that improve America's health care system and respond to other important social needs were available on-line, without waiting in line, when and where you needed them;

You could live in many places without forgoing opportunities for useful and fulfilling employment, by "telecommuting" to your office through an electronic highway instead of by automobile, bus or train;

Small manufacturers could get orders from all over the world electronically—with detailed specifications—in a form that the machines could use to produce the necessary items;

You could see the latest movies, play the hottest video games, or bank and shop from the comfort of your home whenever you chose;

You could obtain government information directly or through local organizations like libraries, apply for and receive government benefits electronically, and get in touch with government officials easily; and

Individual government agencies, businesses and other entities all could exchange information electronically—reducing paperwork and improving service.

In order to implement this vision, the NII's proponents have cited resource needs in four areas: *network infrastructure,* all the apparatus that enables people and machines to swap information between locations; *computers* and information appliances; *resources* such as databases, applications software, and training; and *people* to operate facilities and provide services. Of these, the last two areas seem to show the greatest need for additional investment. Networks and computers, on the other hand, seem in far better shape. Private-sector communications providers—such as AT&T, Sprint, and NYNEX—and corporate communications users presently spend more than $1 billion a week on network improvements in the United States alone. And the computer industry spends at almost twice that rate to meet its needs. Although significant technical issues persist—such as how to comingle audio, video, still images, and data streams of all kinds—spending large amounts of public money for network and computing infrastructure hardly seems a pressing problem.

On the other hand, computer networking services—such as those described earlier in this chapter—favor the "haves" over the "have-nots" in areas such as computer literacy. How many of our "average" citizens can call upon the equipment, training, and support needed to use a system originally designed for Ph.D.'s, after all? Moreover, who will supply the expert physicians and gifted teachers needed to serve millions of electronically linked clients? And who will support them? Only in the industrial-design segment do we see significant resources volunteering—largely from government-supported laboratories in search of new missions.

Before I sound too gloomy, we ought to note that scarcity in human resources translates into job opportunities—a welcome prospect in an era of productivity-driven disemployment. In that connection, my earlier suggestion that virtual science faculties be created at the Department of Energy weapons laboratories provides an example of matching surplus skills with emerging needs in other areas.

Clearly, our nation's medical, educational, and engineering services will benefit from improvements in the ways in which practitioners can communicate with one another. But we needn't stop there. While better information access for physicians, educators, and the like will trickle down to society at large in the form of better services, we needn't limit this aspect of the NII concept to traditional computer users.

Consider industrial competitiveness. High-tech businesses don't exist in a social vacuum. Productivity suffers when an engineer frets because she can't get suitable day care for her children. Costs mount when the financial organization must add extra overhead charges to each project to defray unemployment taxes—or when security must hire extra guards to discourage homeless "visitors."

Industrial competitiveness, therefore, depends on more than just what happens on the production line. A truly healthy economy hinges upon our society's ability to provide meaningful opportunities for all its citizens. Yes, many of today's job seekers with laser-printed résumés face tough times. But consider the plight of those less fortunate. Wait in line at the unemployment office while some clerk thumbs through a mixed bag of leads: *Out of date. Requires a car. Ends too late to put the kids to sleep. Had a different kind of person in mind.* Surely modern technology can offer its citizens more humane access to needed information.

Some years ago, most welfare agencies began providing telephones to families on public assistance as part of a minimum standard of living. Extending that sensible concept into the future, I can imagine a single mother with a school-age child in search of part-time employment making inquiries via her home communicator's language-recognition interface. Having selected "Employment" from a conversational menu, she might ask, "Are there any baby-sitting jobs available during school hours along the number five or twenty-six-A bus routes?" A human–machine dialogue would then ensue as the system searched for and located the appropriate database, then helped explore various options. Having gained agreement on the best match, the system could then put employer and candidate in contact with one another. The first interview might take place later that morning, as the engineer I mentioned earlier and her potential helper conduct a face-to-face dialogue via their respective television sets.

Farfetched? A handful of chips, electronically enhanced telephone lines, and some artfully crafted applications software could do the job. We Americans live in a land in which some one hundred thousand scientists and engineers work for U.S. government laboratories at an annual cost of some $25 billion. Large segments of our industries bewail the lack of customers for their products. Couldn't we engage some of our idle resources on needed services like the one I just described? It seems only natural that opportunities such as these have excited the imaginations of press and public—as well as of our government leaders. But what will it take to transform this rhetoric into reality?

Realizing the NII Promise

Information Superhighway enthusiasts frequently tell the following story to illustrate the social benefit of a national information network. A small-town physician in Appalachia encounters puzzling symptoms in a gravely ill patient. With no medical colleagues at hand, the physician turns to a computer terminal and contacts a specialist at the Mayo Clinic. Correct diagnosis made. Treatment initiated. Patient saved.

While such a story might work well for a few individual cases, how would the consultant in question deal with a steady stream of such requests? In other words, would the infusion of extra communication hardware really solve the problem for large numbers of people?

Rather than debate such an issue in abstract terms, let's look at the state of health care in an area that suffers little from geographic isolation.

Among the world's communities, few can claim anything like the amount of communications technology concentrated on the fourteen-mile strip of volcanic rock known as Manhattan Island. Optical fibers from nearly a dozen suppliers fill the conduits and tunnels beneath its streets, along with massive bundles of telephone wires and coaxial cables. Above ground, microwave radio links and cellular phones provide additional communication capabilities.

Communication aside, Manhattan boasts one of the largest concentrations of teaching hospitals and medical specialists in the entire world. Whatever problems afflict its citizens, distance from doctors doesn't rank high. In particular, take the case of Harlem. In addition to a number of public and privately supported institutions within its borders, the neighborhood includes two of the world's most highly regarded hospitals: Columbia Presbyterian to the north and Mount Sinai Medical Center to the south. With these distinguished institutions and a half-dozen others within a few city blocks, one can hardly blame much of Harlem's medical picture on geographic isolation.

Spared the distance problems that telecommunications enthusiasts seek to address, the state of medical care in this community suffers serious shortcomings nonetheless. Indeed, as an article in the *New England Journal of Medicine* pointed out, few places seem in worse shape from a health perspective. Just a local telephone call, a few minutes' walk, or a short bus ride away, world-renowned medical facilities coexist with death rates so high that adult males in Harlem face shorter life expectancies than they would at the same age in Bangladesh.

Tragic, but hardly newsworthy, it seems. The same issue of the *NEJOM* contained a much cited article questioning the effectiveness of oat bran in lowering cholesterol. The oat bran story received attention everywhere, but the Harlem study got nothing more than a single short article in the *New York Times*.

While the drama of a child saved by a physician hundreds of miles away deserves to strike a responsive chord in any caring human being, self-satisfaction with such an outcome should not blind us to more pressing needs in setting our priorities. As the Harlem example shows, bringing improved medical care to underserved communities requires more than just a data pipeline. Adequate numbers of caregivers and the support staffs required to couple their skills to patients—as well as software capable of holding back a potential flood of extra paperwork—will provide the value that the "highways" in question need to carry.

In this spirit, I hope our society will provide additional human and material resources for our most pressing social needs. As that investment takes hold, improvements in our information infrastructure will stretch existing resources further through gains in efficiency as well as spur the economic growth needed to pay the bill. What role can technology play in bringing this about? Let's take a look at some key elements.

Internet

These days, nearly everyone in a computer-related business owns at least one computer that connects to *Internet:* a group of some thirty "regional" networks that link to thousands of smaller networks and exchange traffic with one another—largely via a common "backbone" network managed under National Science Foundation auspices. At first glance, these regionals serve the same function that companies like Bell Atlantic provide for telephones—while the backbone equates to AT&T or Sprint. But under closer inspection this neat division blurs. Some of the "regionals" serve users from coast to coast and also use other backbone networks to supplement their own transcontinental lines when the government-subsidized backbone doesn't meet their needs.

In almost all cases, Internet carriers lease the lines they use, largely from telephone companies—buying bandwidth at wholesale prices and recovering their costs through a complex maze of fees and government subsidies. These subsidies, in turn, defray the costs charged to schools, government agencies, and various nonprofit organizations. In addition, regionals negotiate a fee with each of their commercial customers, one based upon the bandwidth of the line connecting the customer machine in question to the network. Each region then forwards some of that money to the folks who run the backbones.

Depending on your knowledge (a number of guidebooks on the subject are available) and level of interest, Internet offers lots of interesting stuff. It acts as a wonderfully varied library, allowing users to browse through more than four thousand databases—each of which directs one to long lists of other lists. Directories, catalogs, software, and bulletin boards—you name it, and it probably exists somewhere. A few dozen keystrokes will allow any Internet user to search the stacks of libraries from Syracuse to Singapore via their respective on-line electronic "card indexes," call up images from an art exhibit at one of the world's major museums, or check the structure of complex molecules. Quite a list.

As a global federation of over eleven thousand networks, Internet carries traffic on an enormous variety of subjects for its millions of users. While psychologist Donald Norman was writing an essay on the use of refrigerator doors as family message centers, for instance, he checked on this common American practice in other countries via an Internet discussion group to which he belongs. His query produced enough replies to fill thirty-seven pages of single-spaced text. All these pages can add up—over ten trillion words a year, by one estimate. Quite a difference from the tiny data-networking initiative that got this ball rolling in the early 1970s.

Historically speaking, Internet's role as a national data-networking service received its boost from ARPA (Advanced Research Projects Agency), funded by the U.S. Department of Defense to seed advances in high technology. Thus, in the late 1960s, a forward-looking group of activists created ARPANet as a means of interconnecting America's fledgling computer-science research community via electronic mail.

Once ARPANet got established, researchers at MIT, Harvard, and Raytheon could ship packets of data intended for CalTech, UCLA, and Lockheed to a *data switch* in Boston for shipment to Los Angeles, where a complementary device reversed the process for local distribution.

Primitive by today's standards, this pioneering system nonetheless embodied the *statistical-multiplexing* concept, which distinguishes data networking from voice-based telephony. When two people have a telephone conversation, they generally engage the services of a direct full-time connection, one that lasts until just after the last "good-bye." While some callers may juggle more than one conversation at a time—as with call-waiting, for instance—each telephone call employs its own end-to-end circuit from dial-up to hang-up. Not so with computers. Their exchanges intersperse rapid-fire bursts of activity with long periods of dead silence.

Since data traffic comes in bursts—short *packets* of data separated by periods of inactivity—data switches route packets in much the same way postal workers sort letters. This avoids the need to dedicate a full-time connection to each data call. Like a letter, each packet carries the address of its destination, and each switch along the route uses that address to direct the packet toward its goal.

Unlike telephone switches, which establish a "hard" connection between the end points of a call, data switches act more like post offices. But unlike letters, packets can't travel together in the same mailbag. Digital transmission circuitry can carry only one packet at a time because each packet occupies the full bandwidth of the channel. Furthermore, while data switches generally have equal numbers of input and output lines, no one can predict how many incoming packets will compete for a particular output line at any given time. As a result, each outbound packet must await its turn, queuing up until the channel in question can accommodate it.

Most of the time, traffic flowing into a data switch will average out before any of that switch's queues fills to overflowing. Moreover, when a particular queue fills up too rapidly, some network-management systems automatically generate upstream messages that temporarily refuse traffic earmarked for that queue's destination. Whenever such precautions prove inadequate, however, the unlucky packets in question will get rerouted, or even discarded altogether.

While most Internet users find this network a useful convenience, few would call it flawless—especially for tasks that go beyond swapping e-mailed notes. Messages sometimes get missent, garbled, or delayed. But since some of these faults may originate at the sending or receiving ends rather than inside the network itself, everyday users rarely get flustered. They just try again. On the other hand, tasks such as tying Wal★Mart's point-of-sale information back to its central database system require a much higher level of network performance. So business users normally employ commercial alternatives for their "industrial-grade" needs.

Despite the fact that most users of today's corporate networks employ Internet for at least some of their individual e-mail, they ship more critical material (such as payroll data) via commercial facilities. Nevertheless, Internet's low cost and pervasive connectivity make it an attractive alternative to private-sector offerings whenever users can tolerate the (far from terrible) level of service it provides.

Although it plays a significant role in the lives of millions of technically adept citizens, Internet occupies just a tiny fraction of the world's data-networking capacity. In the mid-1990s, America's commercial data-networking users spent almost $10 billion on such services each year. Compare that to the $100 million spent annually to run Internet in the same time frame. Little more than 1 percent of the total. Moreover, Internet's principal backbone occupies a single forty-five-megabit-per-second digital channel. My best information suggests that approximately one thousand such channels span the North American continent at the present time. Most of this enormous digital transmission capacity moves telephone conversations, but data traffic accounts for a significant—and rapidly growing—share.

Superhighways

When it comes to shipping large quantities of data bits over long distances, nothing beats glass. A hair-thin optical fiber—fitted with a laser blinking on and off billions of times per second at one end, and a detector that converts received light pulses into their electrical counterparts at its other end—can move staggering amounts of information from one location to another. Transparent enough to allow lightwave signals to travel over one hundred kilometers between amplifiers, each meter of fiber costs far less than an ordinary postage stamp. Yet a single strand can provide enough bandwidth to transport the Library of Congress's entire contents to any spot on the globe in just a few hours.

Since a given length of fiber costs a fixed amount, designers of long-haul systems try to raise the rate at which lasers blink on and off to the highest practical levels—especially for ocean-crossing links where cables cost hundreds of millions of dollars. And such demands create all sorts of technical challenges. For instance, semiconductor scientists must devise lasers that yield well-balanced pulses of light every single time, even though they are turned on and off tens of billions of times each second. But none of these device/design challenges exceeds the demands placed upon the fiber itself.

Imagine asking for a material with the following properties: stronger than steel, more transparent than the air outside your window, and less dispersive than any substance known to optical science. Yet today's glass fiber has all these qualities. At the present time, dispersion presents the greatest challenge to fiber designers because it limits the number of different wavelengths that can share a single fiber. As for strength, careful attention to purity and special coatings that prevent the growth of microscopic cracks combine to produce tensile strength far greater than that of the toughest steels. Furthermore, optical amplifiers inside the fiber itself make the cables transparent, as well as tough, so that a detector in France can extract the information contained in light pulses from a laser in New Jersey.

With transparency no longer the prime issue, most current attention has shifted from how *far* one can ship pulses of light to how *fast*. Researchers already know how to switch lasers on and off tens of billions of times a second, but nature always sets a limit someplace—in this case, the fact that light pulses tend to spread apart as they move down a fiber. It's a bit like sending several troops of Boy Scouts on a long hike. Depending upon the spacing between troops and the length of the hike, the slowest hikers in any given bunch will drift back to the troop behind sooner or later—just as the fastest hikers will overtake laggards from the troop ahead of them.

Even though light travels at a constant speed in a vacuum, the energy in a fiber-transmitted pulse splits apart via interaction with the guiding structure, thereby creating this "scout troop" effect. While the laws of physics call for such pulse spreading, scientific ingenuity has found a way around the problem: raise the power in the pulse enough to create a moving, light-confining "well." To see how this works, imagine that each of our troops included a really heavy scoutmaster, one so corpulent that the road bent (momentarily) under his weight—much like walking on a mattress. Under these circumstances, anyone ahead or behind the scout-master would find themselves walking either uphill or downhill and thereby drawn toward a common pace.

This "heavy scoutmaster" phenomenon goes by the name of *solitons* in scientific circles. In nature, oceanographers first observed such bunching in single waves that sometimes remain intact long enough to circle the globe—after an earth tremor, for instance. In communications, the soli-ton approach should allow us to increase the transmission capacity of individual fibers by an order of magnitude over today's systems.

The highest-speed fiber systems presently in service operate at between one and five billion bits per second. With tricks like the soliton effect to look forward to, however, we can expect the continuing pace of technological advance to boost that figure by a factor of ten or more, on the basis of presently understood principles. Even if soliton-based sys-tems prove economically unattractive, fiber-system designers can boost capacity by combining light-wave signals at different wavelengths on a single strand of glass.

Like wires, however, glass fibers suffer from an annoying shortcoming. Both tether their users to fixed locations. So for people on the move—as well as for activities that need to rearrange their communications layouts frequently—wireless technology becomes the pathway of choice. As the U.S. government makes increasing portions of the radio spectrum available for commercial uses via auctions, we can expect cellular telephony to extend toward date networking in tandem with its wired counterpart.

With fibers and radio offering so much technical glitz, everyday copper wires seem something of an anticlimax. Indeed, many people who wouldn't dream of dropping an empty tin can into anything but a recycling bin have urged local telephone companies to scrap their wire-based infrastructure in favor of fiber. But the cost involved causes most people to give the matter additional thought. Estimates of the cost of replacing every pair of copper wires that now links homes, businesses, and other institutions to their local telephone company's facilities range between one-quarter and one-half trillion dollars. Enough money to encourage serious consideration of interim solutions.

The Access Bottleneck

As they move to modernize, America's local telephone companies (and the customers who depend upon their continued financial health) seem caught between the need to maintain revenue from existing voice service on the one hand and the looming threat of competition for data customers from alternate-access providers on the other. What can they do?

With the prospect of greener grass a seemingly irresistible attraction, most of today's cable companies look forward to a role in two-way communication, just as the telephone companies hope to cash in on video entertainment services. Either way, extending fiber lines to large numbers of homes seems a likely part of our eventual communications infrastructure. In the meantime, however—as well as for those cases in which the cost of fiber links may not make economic sense—local telephone companies and their subscribers shouldn't overlook the unused capabilities of their already-paid-for telephone wires.

The wires that link everyday telephones to the phone network do so with the aid of amplifiers that strengthen voice signals. Years ago, the telephone network carried such signals in their original analog form all the way from source to destination. With the advent of digital technology, however, communications designers found it advantageous to convert these analog signals to a digital form as soon as possible, and then to reconvert them to analog just short of their destination. As a result, a phone conversation between my Massachusetts-based son and his sister in California proceeds in the form of a two-way stream of digits for all but a few hundred meters at each end of the three-thousand-mile route between them.

When they want to swap e-mail via their home computers, however, they must do so via modems, which convert digits into tones—only to have the phone company convert those tones back into digits for most of the trip. What a nuisance and what a waste of bandwidth. While telecommunication carriers devote sixty-four thousand bits per second to the digital portion of each phone call's route, most modems can't operate at even a tenth of that speed.

In order to ship data over today's telephone lines, each modem takes the digital pulses it receives at its data interface and converts them into speechlike tones—while also performing the reverse process for traffic flowing in the other direction at the same time. You can actually hear such modems in operation by listening for the chirping sounds that fax machines exchange with one another.

Generally speaking, modems operate at rates of a few thousand bits per second, and there is some hope of their going to about ten times that level in the next several years—almost exactly one-thousandth of the bandwidth we just spoke about in our fiber discussion. No wonder fiber enthusiasts want to scrap wires.

But note that the bandwidth of the *analog* link—which carries speech tones in their original form—comes from electronics tuned to transmit human speech rather than from the wire itself. And this suggests an alternative approach: convert speech into digits inside the telephone itself to eliminate the need for analog transmission altogether. In that way, telephone customers who value digital bandwidth can avoid the analog bottleneck altogether without going to the expense of a fiber link to their desktops.

ISDN (Integrated Services Digital Network) offers a technological bridge for providing end-to-end sixty-four-kilobit-per-second service for now frustrated data users. I recently used an ISDN line to discuss the future of telecommunications with science fiction writer Arthur C. Clarke via my office PC's videophone. Clarke makes his home in Sri Lanka, about nine thousand miles and eleven time zones from New Jersey. Fortunately, telecommuting via a couple of (truly) long-distance telephone circuits suited the meeting's sponsors just fine. Since the PC-based video link required twice the bandwidth of an ordinary digital voice circuit, the video call (sound included) ended up costing just twice the normal rate for a regular overseas phone call.

While almost all homes still use analog telephone service, this all-digital alternative is gaining growing acceptance in business. With ISDN, the telephone itself converts a speaker's analog voice signals into digital form right in the unit, allowing digital service directly to the desktop. Since several of my colleagues now have ISDN at home as well as at their places of work, the wires that connect our phones to New Jersey Bell's switches in these locations provide 144 kilobits per second of digital connectivity—enough bandwidth for two 64-kilobit-per-second channels as well as some extra signaling.

With the prospect of greater bandwidths, some computer makers already supply ISDN ports as standard features, and a growing number of leading-edge PC users are investing in plug-in ISDN boards. Together, that should make a lot of screens fill up faster. But that's not all. As all-digital computer access via ISDN becomes commonplace, applications that make use of this increased bandwidth will flourish. As my Sri Lanka anecdote illustrates, I frequently avoid the hassle associated with travel by presenting speeches over a video link.

Until recently, such arrangements called for elaborate (and expensive) satellite connections. Nowadays, however, I use one of these new low-bit-rate video boards in my personal computer. This system requires only two sixty-four-kilobit-per-second digital channels, available from the ISDN telephone that sits atop my desk. In this way, I can now address groups in Atlanta or Stockholm without leaving my office. On the other end, the conferees normally get hold of a similar unit and arrange for phone lines with the local hotel or conference center.

Since most modern telephone switch designs already provide for ISDN connectivity, installers need merely replace an analog interface board (or *line unit,* as it is called) with its ISDN equivalent and reconnect the subscriber's wires. Operational savings to the telephone company—as well as the opportunity to offer subscribers additional services—ought to make ISDN deployment increasingly attractive from an economic point of view.

Whether ISDN, or some other digital networking style, proves most attractive, the fact remains that the historical evolution of telephony from a totally analog system to a digital network with analog stubs at its ends stands in dire need of improvement. When someone in Atlanta sends a fax to London, that person pays for a sixty-four-kilobit circuit but only gets to use one-tenth of that bandwidth. Imagine sending each page at the present rate, but with ten times the resolution, and all at today's long-distance prices. Only the analog bottleneck at the end stands in the way.

Voice/Data/Video Merger

Opening up the analog bottleneck that now constricts most telephone access lines will provide significant improvements. But we need to move further. In today's environment, voice, data, and video come to us via three separate sets of networks that barely connect with one another. Can't we expect better? Suppose, for example, that you wish to swap images or video in the middle of a telephone call. Right now, that would normally involve separate terminals and connections, so let's see how new technology might improve matters.

While proclamations of a merger between computing and telecommunications date back to the 1970s, technical issues have helped to keep the world split into two networking regimes for decades. So even the early calls for a National Information Superhighway presupposed a separate "telephone" network. But such a split makes far less sense today, thanks to a technological bridge that allows both kinds of traffic to coexist on the same network. As with most such advances, this new technology depends upon a series of earlier inventions.

Merely converting voice traffic into the digital packets we discussed earlier doesn't guarantee good transmission. Unless voice packets get special privileges—such as access to separate, faster-moving queues at switch points, for example—annoying delays can introduce unintended gaps in midsentence. Moreover, heavily loaded packet switches frequently divert incoming traffic along alternate paths, causing packets to arrive out of sequence at their destination and thereby introducing additional delay into the transmission process as they get sorted out.

In the late 1960s, computer scientists began to address the problem of keeping the packets of each message from getting out of sequence with respect to one another. Fortunately, they found an elegant solution—the so-called virtual-circuit concept. Why repeat the full address on every packet in the same message? they reasoned. Instead, they came to understand that the members of such a message group could merely carry a highly abbreviated address, say "BOB," which would designate a prescribed path, one blazed by the first packet in the sequence. Under this scheme, the first packet named "BOB" carries a flag that says to each switch, "Open me up and read what's inside."

In response, each switch in this virtual-circuit network reads the address carried in the trail-blazing packet's message section, sends it in the right direction, *and then sets itself up so as to send all other packets named "BOB" along that same pathway* (thereby creating a virtual circuit). When the calling machine completes the message in question, its local gateway sends one last "BOB," which again requests each switch to "look inside." This time, the message instructs each switch to terminate the arrangement—much as the act of hanging up at the conclusion of a telephone call breaks all the connections between its two end points.

While this virtual-circuit scheme attracted considerable scientific interest—as well as at least one successful commercial data switch—in the years following its invention, most data-network architects continued on the simpler (so-called *datagram*) path described earlier in this chapter. As we've already noted, datagram networks accord each packet much the same individual treatment that a first-class letter receives in the U.S. mail. Over time, however, the datagram approach has become less attractive. Technology has become cheaper, and the demand for voice/data integration larger. And that combination has tipped the scales toward a new look at networking standards.

Today, therefore, the world's computer and telecommunications companies anticipate the advent of a single networking style (called ATM, for *asynchronous transfer mode*), which will accommodate a rich mixture of voice, data, video, and still images—without placing any one modality at a disadvantage. ATM packets, or "cells," as they are called, come in a single rather short length, with limited space for an address, and rely upon the "look-inside" scheme I mentioned above.

What can we expect from this technology? Nothing less than the integration of voice and data traffic in a unified communications infrastructure. As a result, I believe we can soon look forward to the benefits that integrated telecommunications can offer.

For example, WYSIWYG ("What You See Is What You Get") is a much used acronym in desktop publishing. As we merge voice, images, and data, I expect to see a similar term in networking, WISIWYS ("What I See Is What You See")—the visual sharing of information so critical to meaningful interaction. Instead of constantly traveling in order to sit across from one another at some table, future conferees will be able to look at one another without ever leaving their offices. During such conferences they will be able to display information, discuss its features, and modify it interactively because they can experience their colleagues' reactions at the same time. This integrated networking fabric will therefore offer shared visual spaces—together with shared access to data.

We will certainly need to bring numerous software applications into future conversations as well, so the video windows on future computer screens will share that space with other windows—ones that present graphics, documents, and text. All playing together, the combination will provide a worthy vehicle for productive trips down tomorrow's Information Superhighways. And, since this unified network should become as widely available as today's telephony, the benefits of these capabilities will extend beyond business to serve the personal needs of individual citizens as well.

Instead of linking to individual telephones or TV sets, tomorrow's networking providers will likely connect to a data switch located on each individual subscriber's premises. These switches, in turn, will tie all of that home's appliances to one another as well as to the outside world— furnace, air conditioning, alarm system, lighting, microwave oven, you name it. Much as today's answering machines and faxes plug in to standard telephone outlets, all of tomorrow's appliances will connect to a full-fledged *home network*. The Information Highway, in other words, will also make house calls.

As government, industry, and concerned citizens address the social, technical, and economic issues that surround the National Information Infrastructure promise, I hope that all of us will keep a straightforward question in mind: Does the system actually serve all its intended users? In that spirit, I believe that considerations of interoperability, ease of use, and environmental impact can help guide decision making on behalf of all networking users—from individual consumers to the largest international corporations.

CHAPTER 7

USABILITY

Technology plays a dual role in modern life—sustaining it on the one hand but complicating it on the other. To paraphrase psychologist Don Norman, author of *The Design of Everyday Things*, we have come to depend upon things that make us feel stupid. And a high-tech education doesn't provide relief from this dilemma. I can't tell you how often professional computer scientists rewrite software from scratch rather than reuse similar codes of their own creation. Too hard to figure out their own program's inner workings.

Why would engineers design things that even they themselves can't use? In exploring this paradox we must contend with an apparently durable attribute of human nature. Life appears to reward action over explanation—the rugged pioneer rather than whoever cleans up the details afterward. In earlier times, this priority forced automobile dealers and their customers to "clean up" manufacturing oversights. While attention to quality has reduced such defects sharply, other flaws remain, and they lie deeper. Many stem from a lack of consideration for the value of convenience to the ultimate user.

The lack of focus on the customer that characterized paperwork-era organizations extends beyond the world of corporations and government agencies. In particular, bureaucratic behavior remains alive and well in settings we don't normally recognize as bureaucracies. Consider the following anecdote from Norman's *The Design of Everyday Things*.

In its role as the U.S. government's "landlord," the General Services Administration received a pair of nearly simultaneous requests from the Federal Aviation Administration. FAA staff in two West Coast cities—

Seattle and Los Angeles—needed new quarters, thereby presenting the GSA with an opportunity to conduct a side-by-side experiment.

While the Los Angeles project proceeded normally, the Seattle building's architect first interviewed prospective occupants and incorporated their recommendations into the design. Not surprisingly, the Seattle tenants liked their quarters better than did the folks in Los Angeles. While job satisfaction rose seven percentage points as a result of the move in the former city, it stayed flat in the latter. Taxpayers felt the same way, it seems. Like the FAA staff, visitors who were interviewed gave the Seattle building higher marks, especially because it didn't look to them like a typical federal office building.

But the story doesn't end there. Can you guess which building won prizes for its architects? The Seattle design "lacked discipline," one award committee noted, thereby defending its selection of the Los Angeles edifice as an embodiment of architectural excellence. Never mind the wishes of the people who used it daily. The professionals, in this case, felt they knew more about what the public ought to prefer than the public did.

An isolated incident? Hardly. Using everyday examples ranging from faucets to instructions for taking medication, Professor Norman's book describes the artifacts of a world in which experts strive for recognition from other experts.

Isaac Bashevis Singer captured this attitude in a response he gave to an often asked question. Starting out as an obscure writer in an obscure language, Singer wrote exclusively in his native Yiddish. While translations of his work would lead to worldwide recognition in the 1970s, his earlier readership was largely restricted to the dwindling part of the American Jewish community that still reads publications in Yiddish.

In order to widen his audience, Singer trod the community-center lecture circuit in the late fifties and early sixties, coming to my hometown some fifteen years before the two of us shared a Stockholm platform to receive a Nobel Prize. At the end of an entertaining lecture, delivered in fluent English, a questioner asked, "Why do you write in Yiddish, when so few people can read it?"

"I once met a man," Singer replied, "who had just written a book in Aramaic. 'Why did you write in Aramaic, when so few people can read it?' I asked him. Do you know what he told me? 'I don't write for the ignorant,' he said."

Singer went on to explain that his choice of his native language as his medium came from the connection between the dreams and demons of his imagination and the words attached to them in the childhood landscape he wrote about. While Singer needed the dying language of another time and place in order to reach modern readers, his Aramaic-bound colleague reveled in obscurity. An expert, out to impress only his fellow experts.

How do these examples relate to our harmony concept? As long as experts focus on impressing their fellow experts, Information Age technology and its users will remain at odds with one another. Sure, technologically challenging problems demand sophisticated solutions. But truly high technology needn't burden its users with awkward complexity. Think of the awesome tools nature bestows upon living creatures. When a German shepherd explores the environs of a trash can, ultraprecise microchemical analysis of organic molecules kicks in automatically. The pooch in question needs only to sniff.

While humanity's technology can't hope to match nature's, the example of the latter seems clear. Truly sophisticated technology should provide its users with unself-conscious empowerment. For this reason, successful interfaces between humans and their information-processing tools ought to mimic accustomed behavior as much as possible.

"When will we be able to talk to our computers?" That question has popped up at the conclusion of my "Future-of-Technology" talks ever since I began speaking about such matters in the mid-1970s. Like most technologists in those days, I generally tried to persuade my audiences that merely replacing typed commands with their spoken equivalents on a one-for-one basis would fail to improve matters. While technically correct, however, my argument missed an important point. Natural language demands usability. Can you imagine anyone having to say "Alt F5," for instance? In spoken language, the burden for accommodating to an interface shifts from the human user to the machine in question or, more properly, to the software that drives it.

For all the hoopla about easy-to-use interfaces, systems that operate on the basis of button pushes generally require users to push the "right" buttons, in the "right" sequence. Natural-language front ends, on the other hand, rarely impose such rigid constraints on their users' behavior. As a result, computers that listen for commands must work harder than those that only need to read typed instructions.

Given a choice between speaking and writing, most people would prefer to express their wishes verbally—just as when taking information in, most would prefer the greater clarity and pace control that written text offers. Since we expect humans to occupy the role of "boss" in dealing with their electronic "subordinates," I expect users to favor a machine that they can command via natural-language inputs and that will respond with machine-created text and graphics. For example, suppose you used a voice-based system to arrange a plane trip. I'm pretty sure you would want a text-based summary of your itinerary. I know I would.

But no one approach will suit all applications. Imagine trying to operate a piano by voice commands, for example. Different interfaces lend themselves to specific task areas. In that spirit, this chapter will focus on the role of technology in the three key domains of direct information access—people with people, people with machines, and people with data.

Among these, the people-with-people aspect emphasizes the creation of shared visual spaces—electronic "windows" through which participants can see one another as well as all items of mutual interest. Accordingly, image-processing software plays a key role here. In addition, the technology involved must support smooth conversion from one communications medium to another—such as the option to listen to selected portions of your electronic mail over a telephone. In such an environment, simultaneous communication via multiple media should present users with no more difficulty than present-day voice-only calls.

Moving to communication between people and machines, we see digital circuits controlling machinery of all kinds—from power plants to cardiac pacemakers. Getting machinery to comply with human wishes, therefore, boils down to expressing those wishes in terms that a machine's control computer can understand.

Over the years, computer instructions have evolved from cryptic, small-scale commands toward more powerful constructs that bear at least a rudimentary resemblance to natural language. Obviously, users will welcome the ability to express instructions in the far less restricted vocabulary and grammar employed in everyday discourse.

But even between humans, speech alone rarely suffices. Imagine Julia Child attempting to teach cooking by sitting behind a newscaster's desk and just talking viewers through the process of creating chicken Marengo. In such an instance, one gesture would be worth a thousand words. Since gesture-based inputs remain a vital aspect of such tasks as playing a piano or steering an automobile, a similar capability is emerging in the computer arena as well.

On the people-with-data front, system designers face an important challenge. In addition to providing access to the massive amounts of tabulated data a connected world will inevitably produce, they must give users improved means of mining information nuggets from unstructured data—all the stuff sitting in billing records or customer complaints, for instance. While far from solved, this task has become less daunting through the growing use of the *intelligent agents* I mentioned earlier—programs that derive and reshape data structures so as to make their contents more readily understandable. Here again, I see an important role for computer-enhanced processing of natural language—especially as we busy humans set our intelligent agents to the task of plowing through all the reading material we never have time for on our own.

Let's look at each of these three themes in more detail. I hope the stories and examples that follow will help bridge some of the gaps in understanding now separating technology creators from technology users. After all, we can expect such distinctions to diminish as people become more comfortable with their command of technology.

People with People

Beneath the forbidding interfaces that computers often present to users, we can find lots of opportunities for enhancing communication between people. Over time, we should look forward to using computers in a conversation with no more hesitation than, say, showing someone a photograph or making a rough pencil sketch. While computers can enhance face-to-face meetings by acting as information providers, I expect far greater value from their role in supplementing tomorrow's "telephone" conversations with other communications media—such as video, graphics, text, and shared software. That way, a pair of financial analysts located in different cities might discuss a problem in a face-to-face manner while working from the same spreadsheet. Thanks to advances in video compression, we can now ship all sorts of television material in a tiny fraction of the transmission bandwidth such signals previously demanded—thereby making video communications a likely commodity by the late 1990s.

With increasingly easy access to a wide range of programming, video telephony, and other forms of image-based communication, today's television screens will evolve into multifaceted communications windows. But I don't expect this communications progress to stop with just the display of images. In particular, I expect future technology to enhance telecommunication with a convenient means of manipulating everyday objects at distant locations. In order to accomplish this "trick," image-processing circuitry would enable remotely located video cameras to act as extra pairs of "eyes" that work on behalf of a system's human users. Still in its infancy, so-called computer vision enables some systems to identify image elements and put them together in various combinations so as to match template examples stored in memory—a useful first step toward the processes that each living being's eyes and brain accomplish with such unself-conscious ease.

Once a computer recognizes an object and can track its progress through space in one location, it has enough information (at least in principle) to instruct an electronic partner to replicate that same motion with a similar object at another location. Instead of an exhaustive series of typed commands, therefore, a system user could manipulate objects at a remote site by merely performing the desired task locally. In one such scenario, friends might play cards with one another while separated by thousands of miles—with nary a keyboard or display in sight. No screens, no typing, just shuffling and dealing an ordinary deck.

In the Harmony Era, I would expect individuals and families to enjoy more natural means of maintaining ties in the face of physical separation, something better than merely exchanging a postcard or phone call. For instance, a twenty-first-century move from Seattle to South Bend might not need to interrupt a long-standing series of Thursday-night bridge games between two couples who really value each other's company. Time zone differences will require some schedule adjustments, but a one-hour shift on each end doesn't seem unworkable.

As for the game itself, some folks might settle for a straightforward extension of a 1990s computer-bridge game. The computer program shuffles and deals electronically, while each of the four participants sits in front of a screen and plays the hand by typing on a keyboard. Okay for bridge freaks, I suppose—but hardly likely to engage anyone who cares more about the people involved than the mechanisms of the game.

Instead, I see an ordinary bridge table, complete with real playing cards. Two players sit comfortably at their assigned places, and two easels stand at the remaining sites. Each easel holds a flat video display—the size and shape of a framed oil painting, say—that shows a head-to-waist image of the corresponding player at the other location. Except for a small box that resembles a two-slice pop-up toaster, all else looks and feels as it did in preelectronic days.

At the beginning of the hand, a Seattle dealer might pick up a deck, shuffle normally, and deal around the table—with the two slots in the "toaster" getting the cards intended for the folks in South Bend. Having put their own deck in the "toaster" in South Bend, the hands dealt to them in Seattle "pop out" onto their table at the conclusion of the deal. As play ensues, the "toaster" at each end supplies the cards played by humans at the other table. Humans collect the tricks, comment on each others' play, ask "Who dealt this mess?" and generally enjoy themselves.

How do the toasters "know" which cards to play? Simple. Just put a magnetic stripe on the back of each card. That way, the toaster's "reader" can tell them apart, but humans merely see the same black stripe as part of each card's design. During play, a video camera reads each card as it appears faceup on one table and promptly instructs the toaster at the other end to supply that same card to the playing surface of its table. For all I know, deluxe models of such a system may feature a separate dispenser that sits at the bottom of each easel, so that those cards will come flying out from below that player's image.

Note the difference between today's seemingly high-tech computer-bridge interfaces and the almost natural environment provided by the real-cards interface. To many people, the first one connotes "technology." No cards. Everyone typing at a terminal. Lots of "help" from the computer involved.

But technology needn't require obtrusive machinery to be helpful. As this example shows, truly high technology should allow humans to act in exactly those ways that suit their unconstrained preferences. Why should living with machines force us to imitate our servants, after all.

Since I haven't played bridge since my college days, I'd be more likely to use such a system to play Go Fish with one of my grandchildren. A computer screen with a video window might let us look at one another as we play, but I'd much prefer to pass real cards back and forth. While I'm sure that some clever programmer will come up with software that allows five-year-olds a chance to cheat once in a while, I'd like my granddaughter to be able to peek at the top card in the pile before she decides what to do.

This playing-card example introduces a larger issue. Exchanging mail, speaking over a telephone, seeing one another over video, and interacting in the shared visual space of a display screen offer us increasing degrees of contact over long distances. But our technology vision needn't stop there. Remote mechanical actuators, like the toasters described above, can respond to remote commands, after all. Moreover, commands needn't come from keyboards. Instead, a truly usable machine should infer its employers' intentions by observing their actions.

Consider the tragic waste too often associated with social mobility in our society. While some healthy retirees enjoy golf and swimming in warm climates, the later years of life often mean lonely days in nursing homes for others. In one part of the world an elderly lady sits in her wheelchair, filling the time until supper as best she can, while in another city her son tries to make dinner and deal with the needs of a first-grader who'd like some attention. Wouldn't all three welcome the prospect of a few hands of Go Fish between grandmother and grandchild?

Instead of worried weekly telephone calls and twice-a-year visits to reintroduce the kids to an elderly stranger, technology can help the generations to stay in touch. Someone once observed that "grandchildren and grandparents get along so well because they share a common adversary." The human race has enjoyed the bittersweet benefits of extended families too long to sacrifice them for the sake of the so-called rat race.

In addition to strengthening emotional ties, technology can offer material benefits as well. While we can surely look forward to more flexible mechanical aids for handicapped people, some infirmities—such as uncertain muscle control or speech impairment—will make it difficult to operate machinery. Under such circumstances, a remote visitor might supplement the local nursing staff in some situations.

Imagine "sitting down to dinner" with an elderly relative in another city. Instead of calling an attendant to fetch a sweater, open plate lids, and the like, the visitor could guide on-the-spot machines through those chores as the meal proceeded. No matter how competent the institution or sophisticated its technology, the extra care provided by concerned human beings for one another adds the missing ingredients needed to make the experience a family meal.

At its most basic level, the underlying technology still consists of ones and zeros—data flying back and forth between slivers of silicon connected by threads of glass. But look at what that technology can do. Or, more properly, what humans can do with it. The choice is ours, after all. And in this application, at least, technology can help future generations soften the impact of physical separation.

People with Machines

When it comes to understanding all but the most straightforward of human wishes, computers still fall far short. Back in the days when the world's most powerful computers barely performed as well as a modern PC, pundits proclaimed the advent of "artificial intelligence." After a while, they even capitalized the term. But what really happened? Lots of interesting ideas, important progress in related fields, and some humbling insights into the awesome subtleties of the thinking engines that humans carry between their ears.

Had Artificial Intelligence achieved its originally stated goals, no one but computer scientists would need to think about software anymore. Machines would simply program themselves. Instead, however, modern reality confronts us with programming chores every time we need to punch more than a single button to accomplish a task. Just look at VCRs. Thousands of these familiar contraptions continue to blink "12:00" around the clock because their owners haven't found the sequence of button pushes that would set them properly.

Despite AI's failure to produce brainlike computers, however, mainstream computer science owes these zealous researchers a considerable debt of gratitude. In particular, because AI researchers helped produce many of our present-day techniques for extracting meaning from multiple databases, we should acknowledge AI's role in providing the world with computers that act on their human users' behalf. Since these so-called intelligent agents "understand" what humans say in only a very restricted sense of that word, language-understanding software offers us our most promising path to hassle-free human–machine interfaces—bridging the gap between the precisely structured directions such agents require and the looser formats preferred by nonexpert users.

Before a computer can parse sentences and extract their meaning, however, speech recognition software must first convert sounds into text. Speech processing, as we know it today, got its start in the mid-1960s when it became advantageous to send the *numerical description* of a signal down a communications pathway, rather than the signal itself. Rather like the post office sending a cookie recipe in place of the cookies, it involves a trade-off. We no longer have to worry about breakage en route, but the resulting need to analyze ingredients at one end in order to bake a fresh batch at the other seems like a lot of trouble.

While the trade-off involved in sending a digital replica may not make much sense for baked goods, digitization swept telecommunications some time ago and continues to make inroads in other areas as well. For example, CDs contain nothing but digits—long lists of numbers that trace the pressure changes on some musician's microphone. *Records,* disks whose grooves track sound-wave motion directly, occupy less and less space in what my generation still refers to as "record stores." The digital revolution has certainly taken hold.

At the lowest level, speech recognition software normally runs on microprocessorlike devices called *digital signal processors,* or DSPs. In order to break apart and analyze the waveforms created by human speech, the DSP's recognition program analyzes speech's spectral patterns to extract groups of so-called *phonemes,* which constitute a kind of speech alphabet. Phonemes combine to create syllables, just as syllables combine to form words—and so on up the chain of meaningful expression.

Rapid progress in DSP technology, and in the algorithms that direct this handy hardware, have combined to make "computers you can talk to" an emerging reality. For example, an exhibition at Seville 1992 Expo featured an automatic language translator that allowed two people—one speaking only English and the other only Spanish—to conduct a real-time currency exchange transaction.

The "translator" in question consisted of a group of language-processing programs—speech recognition, language understanding, translation, and text-to-speech conversion, running on a special speech-processing circuit board. This board occupied a plug-in slot of the high-performance workstation that acted as its landlord—providing power, microphone and loudspeaker connections, data storage, and the other housekeeping services that computers normally supply. That's all. Yet it managed to sustain a real-time conversation between two people speaking about a common subject in two different languages.

While this system's designers achieved a significant milestone in interpersonal communications, the "translator" also marked an important step toward computer usability. In particular, we can view the translation system as a pair of human-to-machine system interfaces backed up against one another. From that perspective, the English subsystem demonstrated its ability to understand speech via the outputs of its Spanish counterpart, and vice versa.

Limitations in the number of prerecorded human speech samples (against which this particular system could match its users' input), restricted the vocabulary to a few hundred words, but continuing progress has since provided us with even greater capabilities. In particular, new speech-processing techniques now permit larger vocabulary sizes, while advances in linguistics promise to broaden the areas of discourse.

On the linguistic side, language ambiguities continue to pose formidable obstacles to full-scale discussion. For now, language-understanding researchers generally finesse this problem by restricting their systems to specific application domains, like currency exchange. After all, a pair of human speakers sometimes talk past one other until they realize that each has a different context in mind. Until they find out otherwise, the different patterns each hearer extracts from the same set of words sometimes lead to confusion, followed by embarrassed chuckles when the participants realize their mistake. Since computers offer only limited help in situations that call for common sense, we need to guide these dumb-but-diligent machines by specifying the context in which we wish to engage their services.

People with Data

Normally, we think of computers as dealing with formatted data—such as the U.S. Census Bureau's list of our country's inhabitants, their addresses, ages, and similar attributes. And most of today's conventional data processors follow such a model. A program that resides on a typical data-processing computer might ride herd on a number of databases, pulling pieces into the computer's processing engine for examination or updating. However, I see two avenues for improvement: an easier way of gleaning needed information from a large number of databases as well as ways of making individual databases more helpful. Owing to the speed and efficiency with which mainframe computers accomplish their traditional tasks, a so-called *master–slave* approach has dominated data processing since its earliest days. Even today, despite all the talk about the mainframe's demise, large corporations (including some that sell alternative solutions, by the way) still buy billions of dollars' worth of mainframe computers every year for certain large operations—such as payrolls.

In such a system, a large number of payroll clerks might sit in front of their (slave) terminals and interact with the mainframe by means of a prescribed set of commands. Once trained in the system's ground rules, the clerks can pursue their chores with little difficulty. Not so for casual users, however. Say that a human resources manager needed to query the payroll database some weekend, with no clerks on duty. Working from home, she might dial up the mainframe in question, get her PC to pretend to be a terminal, and query the mainframe in much the same way as the payroll clerks do during normal working hours—provided, of course, that she had all the proper procedures at her fingertips. While she might solve that problem with a phone call or two, what about databases further from home?

In today's networked world, increasing numbers of computer users, such as this HR manager, need to act as casual users of databases around the world. Unlike full-time users with special training, they can't keep track of all the individual idiosyncrasies, data structures, naming conventions, and addressing layouts of the machines involved. Fortunately, new interface software now offers them more democratic access to such facilities.

In describing one example of this new capability, called Mosaic, the *New York Times* asked its readers to think of it as "a map to the buried treasure of the information age." This highly popular program stems from the work of a small group of software developers and students at the National Center for Supercomputing Applications in Champaign, Illinois. It is available free to Internet users, and the center downloaded several hundred thousand electronic copies of it to individuals requesting it in the first year alone—and received more than six hundred thousand electronic information queries *per week* from Mosaic users.

Mosaic's list of commercial users ranges from traditional software houses like Lotus and Novell on the one hand to catalog distributors and record companies on the other—not to mention students, hobbyists, librarians, and anyone else with a yen for exploring the world's data highways.

What makes Mosaic so popular? Imaginative use of the so-called *client–server* model of computing. In this environment, software on the user's machine engages various targeted database-computers in dialogue on that user's behalf—thereby shielding that person from their respective inner workings. Instead of wrestling with arcane commands, users merely click on interesting menu items displayed in windows on their local computer screens. And so, as commercial software companies adopt the Mosaic platform to create new data-exploration products and services, use and usability will surely grow.

In addition to programs—such as Mosaic—that simplify access to existing databases, computer science also offers the prospect of help from individual databases themselves. Alongside tabulated data, these *active databases*, as they are called, might contain software that monitors data on behalf of a particular client and takes prescribed action. In order to see how this might apply to a hypothetical case, let's consider a computerized version of the stock market tables in your favorite newspaper. Suppose you dabbled in the stock of a budget motel chain that catered to vacationers. In that case, you might wish to track trends in domestic versus overseas travel so as to time your trading accordingly. If more Americans decided on cross-county tours in preference to overseas vacations, one might expect more use of the domestic motel chain. One might also expect more sales of new autos and fewer reservations on international airlines. On this basis, you might tie your stock purchases and sales to simultaneous moves in auto and airline share prices.

Imagine that you wanted your computer to alert you just as GM and British Airways stocks moved more than a given amount (in opposite directions, of course) in three consecutive sessions. Not too hard, provided that your computer didn't have too many other obligations. Just instruct it to query the on-line data every few minutes and see what's happening. On the other hand, if the number of such obligations grows appreciably, as in the case of a brokerage firm offering this alerting service to its customers, the machine might easily run out of steam.

Active databases to the rescue. With this technology, a tiny program associated with the database itself monitors the flow of updates into the stock tables, keeps track of changes in GM and BA as they occur, and only troubles the main computer when a sequence of events triggers its preset value. Rather than merely acting as the main computer's passive servant, the airline database reports on itself—thereby avoiding the huge number of fruitless inquiries that the main computer would otherwise make in order to remain abreast of the situation.

While I used an ordinary stock table in this example for the sake of simplicity, I would expect much of tomorrow's data to reside in less structured formats, which would benefit from an *object-oriented* approach. Imagine a "listing" of companies that not only included such items as magazine articles about them, product brochures, and even video clips of news stories, but also responded to queries about their content. By associating specialized operations with each data type—such as image analysis with an illustrated brochure—the object-oriented approach can allow questions such as "Which keyboards have a tracker ball located just below the space bar?" if the catalog in question contained pictures of laptop computers, for instance.

In addition to accommodating a rich variety of data types and the operations that we might wish to perform upon them, object-oriented databases make tracing relationships much easier than do conventional systems designed to scan tables of numbers. Just as an object in an object-oriented database contains all the operations one might perform upon it, it can also contain pointers to all the other objects with which it relates—such as a workstation's assigned user, sales representative, repair service, and software packages, for instance.

With conventional databases, a corporation's Management Information Systems organization might list the equipment it owns in one file, service records in another, and software purchases in a third. Imagine tracking vendor performance on the basis of such widely scattered information. Several so-called *value-added resellers*, or VARs, might offer similar graphics workstations, for instance. Which one does the best all-around job of hardware/software integration, testing, and support? With each workstation tabulated as an object, the information needed to answer such questions becomes available in a single package.

As the ability to trace relationships becomes more important, this new technology can also make life easier for database users by keeping all members of a data "family" aware of any changes in their relative's status. If, for example, a repair service changes subcontractors, a new entry in just one place—the object that represents that contact—will produce up-to-date answers to service-related queries throughout the database.

From a user perspective, therefore, we can look forward to richer and more timely information from databases. But that's only half the story. Most information lies outside of the formally organized structures we call databases, after all. Suppose someone offered to double the flow of unformatted information to your home or office, such as those magazines you always mean to read. Would you welcome that prospect? Right now, I'd expect most people to say, "What would I do with more information? I'm drowning in piles of paper already."

In most offices—including my own—every horizontal surface seems to have its own stack of memos, magazines, and other papers. Without diligent pruning, that "foliage" quickly spreads beyond manageability. Why would any sensible person welcome the prospect of increased information flow in such an environment?

Let's examine this issue. Wouldn't more of the right information make you more productive? If any of us really possessed all the information we needed, we could toss all those unsightly stacks into the nearest recycling bin—and save a tree or two in the process. Why don't we? Because these heaps of extraneous verbiage also contain nuggets we need to know about. We don't dare let them drift out of reach. It seems to me, therefore, that people generally have messier offices when their ability to access information fails to match their needs.

But the ways in which we employ paper, rather than that medium itself, determine its usability. When you think about it, the retrieval mechanism underlying those much maligned stacks of paper has much to recommend it as an information-storage system. Why else would it flourish? As you search downward, you first encounter the items you added (or put back) most recently. Thus, the material you tend to use most frequently remains near the top, while the less useful stuff tends to move downward out of the way. But every convenience exacts its price. Since no one wants to sort through hundreds of items, a tall stack tends to multiply into shorter ones. It's easier to get to the bottom of a small pile—provided you're searching through the right one.

With conventional information pathways clogged by paper overload, eager entrepreneurs touted electronic alternatives as early as the mid-1970s. While the notion that those text-based "paperless offices" could replace all paper seems laughable today, information-retrieval systems based upon text files do fill important niches in today's workplaces. For example, most lawyers search for legal precedents via keyword queries to on-line database services aimed at the legal profession.

Clearly, such systems offer advantages over poring through rows of dusty volumes in a law library. Lawyers, or at least their paralegal assistants, can afford to spend whatever effort it takes to mine the contents of a well-defined body of data. They profit from replacing a highly labor-intensive task with a somewhat less onerous one.

Less specialized users, on the other hand, have fewer resources to devote to searching and larger areas to sift through. Even if the computer-adept manager of a small employment agency has electronic copies of all résumés on file, she can't be sure that a particular job title will snare all the candidates qualified for a particular opening. As a result, we can expect her to widen the search by including synonynms in her query.

But what about job seekers—like the mother we cited in our earlier discussion on networking applications? Here, technology must resolve any ambiguity, not the user. Everyday language, after all, does a satisfactory job of conveying meaning. Otherwise, you wouldn't be reading this book.

Despite all the ambiguities that lie within spoken language, written text, images, diagrams, videotapes, and all the other sources of everyday information, the fact remains that humans generally manage to make sense out of their contents. So, while designers of computer-based substitutes for manual data retrieval deserve some sympathy, they can hardly claim their tasks to be impossible. Enhanced through a combination of ingenious insights, diligent engineering, and ever more powerful electronics, therefore, technology should offer more usable access to all kinds of data—irrespective of its format—as well as to people and machines.

In the end, technology vendors will benefit at least as much as technology users. As we have noted throughout this book, today's most promising new business opportunities lie outside existing markets. To reach these new markets, vendors of sophisticated technology cannot aim their products at sophisticated users alone. Just look at the widespread acceptance of pagers and fax machines. As easy to use as ordinary telephones, yet enormously more complex, they rarely make their users feel inept. Imagine what would have happened to sales had these computer-based devices demanded computer literacy from their users.

Like faxes and pagers, tomorrow's blockbuster hits in the marketplace must engage customers who don't relish wordy instruction books. And—to complete the picture—as these easier-to-use products link more effectively with their users, they will need improved links to one another as well.

CHAPTER 8

INTEGRATION

In an integrated world, your electronic helpers can get whatever information they need to accomplish their assigned tasks. No more hunting for receipts at tax-return time, for instance. Think how much hassle this could save us in our jobs as well as in the personal errands that consume much of so-called leisure time.

Earlier in this book I described the transition from quantity-focused mechanical manufacture to the high-quality output of computer-controlled production lines. Information sharing between one machine and another underlies much of the difference. Before the advent of digital controllers that could converse in ones and zeros, people scrambled among deaf-and-dumb machines, working to smooth inevitable glitches. There has been much improvement since those days, in offices as well as factories.

But the ability to "speak" a common language doesn't guarantee a "conversation." Take the state of technology at a typical modern hospital. Every nook and cranny seems crammed with exotic-looking apparatus. And every machine of any size features at least one computer screen and keyboard. With computers everywhere, why do so many doctors, nurses, and other hospital employees spend so much time dealing with piles of barely legible paper?

During a recent hospital stay, I noted that the nursing station in my ward employed a full-time "secretary" who spent most of her full-time workday punching holes in 8½-by-11 sheets of paper and putting them into binders—one for each patient. My own binder accompanied me as I was wheeled from one test to another, and it grew thicker in the process.

As far as I could tell, each of the expensive machines that probed my condition communicated its output via hand-carried notes.

An isolated incident? Hardly. In numerous public forums, as well as in private conversations, health-care professionals confirm this personal experience with the system. Everyone involved bemoans the human and economic waste that stems from the need to bridge gaps between technological islands. And, needless to say, this wastefully fragmented state of medical care differs little from that of other technology-intensive sectors of our society. Despite the proliferation of computers, it seems, systems integration is still little more than a buzzword.

What images does the term "systems integration" conjure up in your mind? Darth Vader's screenful of Imperial battleships gliding through the galaxy in search of Luke Skywalker? In the business world, a Management Information Systems (MIS) manager might imagine an information environment in which every single hardware and software component works perfectly with all the others—much like a well-drilled symphony orchestra. While this still lies beyond the grasp of today's corporate MIS architects, I expect even more from the Harmony Era. In my view, the harmony concept calls for providing that same level of coherence to individual consumers.

Yes, businesses must align their information-shaping and -sharing processes. But such an alignment hardly represents an end in itself. Businesses, after all, exist to serve society. So manufacturers of CAT scanners or cardiac monitors could tune all their internal information systems to seamless perfection yet still fail to meet this requirement. Today's medical technology saves countless lives, but the high price of medical care pushes its benefits beyond the reach of many people. And part of the blame goes to the cost of moving information between machines lacking networking capabilities.

Before we condemn equipment producers, however, we must remember that the markets they serve have offered little encouragement for systems integration. Each of the individual specialists making purchasing decisions has had a specific task in mind—and has rarely seen benefit from the added complexity of matching information standards set by others. In the mainframe era, after all, central computers did little to reduce paperwork, and even spurred its growth in many cases.

On this basis, readers might well picture MIS managers as unthinking troglodytes, their imaginations stunted by overinfatuation with their mainframes. But put yourself in their shoes. Imagine bearing the responsibility of keeping a centralized information system running. Hundreds of thousands of lines of code. Each software fix, or improvement, creates the possibility of subtle interdependencies. Might not a simple change in our line of code or data-collection interface create unforeseen consequences elsewhere? As a result, most of yesterday's centralized systems set standards that left little room for flexibility and discouraged interconnection.

In my own case, I frequently ask for "minor" improvements in the computer system that supports my office—only to find a problem caused by the change. In our relatively small environment, the problems generally prove easy to fix and do no harm to others. But think of the same situation in a modern hospital. Small wonder that systems-integration advocates face a tough selling job, in addition to all the technical issues involved.

Fortunately, modular architecture can make interdependencies more traceable and tractable. While modularity "wastes" space—like placing each piece of dinnerware in its own box when packing the set into a large carton, for example—the value of an electronic "container" no longer looms large when compared to the value of its contents. Given the plummeting cost of microcomputers, we can expect system architects to employ dedicated computers to accommodate the networking needs of a wide variety of machines—leaving the existing system free to chug away in its accustomed mode of operation.

In tomorrow's hospitals, for instance, a networked layer of interface software (called *middleware* by some) must sit between individual units of diagnostic equipment and the system that maintains hospital records. This program would call upon the services of a separate computer module to fetch all information about a particular patient, thereby enabling doctors and nurses to spend less time with paper and more with people.

As computing and communications merge into a single entity, cooperation between specialized systems will surely become more commonplace. And not just in hospitals. Since most people find the increased convenience of systems integration attractive, we can expect escalating levels of complexity in all facets of everyday life. In the Harmony Era, I believe, homeowners will expect their TVs, VCRs, camcorders, telephones, burglar alarms, heating systems, stereos, electronic encyclopedias, personal computers—all their home appliances—to communicate seamlessly with one another. And as these connections increase, the number of possible combinations will grow enormously.

In Chapter 7, we saw how the use of *object-oriented* databases eased the difficulties inherent in dealing with a wide variety of data types. Similarly, the use of *object-oriented* programming reduces the need to anticipate unforeseen encounters between objects and the operations that govern their behavior. By tying expected behaviors to the object in question rather than to the system as a whole, programs employing this new approach no longer require global modification every time a new object—such as a new billing arrangement—appears. In the past, as global rules piled atop one another under more conventional schemes, so did the chances for unforeseen conflicts.

While the number of choices offered by everyday life might appear daunting from a mathematical point of view, we seem to manage somehow. Psychologists who study human decision making tell us that we rely on accustomed paths. Unlike programs based upon global rules, human intelligence rarely seeks global optima. Instead, we normally stop searching when we reach a satisfactory, rather than optimal, solution to a given problem.

In his landmark book, *Between Parent and Child,* psychologist Chaim Ginott advises parents to reinforce their children's associations of objects with their intended uses. For example, he prefers "Chairs are for sitting" to "Don't stand on that chair." While the latter addresses a single situation, the former builds upon healthy human learning. Nature accustoms us to link objects to the roles they normally play in our lives. In that regard, I sometimes imagine holding a hammer in my hand just as soon as I decide to hang a picture.

How does this use of *object-oriented* (as opposed to global) constructs relate to software? In the conventional programming style, multiplying two numbers might proceed as follows:

Let $a = 5$
Let $b = 6$
Let $c = a \times b$

In other words: "Load *five* in one register, *six* in another, and then tell the multiplier to process the contents of those two registers. The correct answer is that $c = 30$. Not bad for numerals, but today's software must process all sorts of other items—from phone calls to elevators.

Even the "simple" act of multiplication depends upon the types of quantities involved. Computing the torque exerted by a wrench involves the direction of the turning force with respect to the handle, for example. Also, while 32,000 and 3.2×10^3 represent the same quantity, these two representations obey quite different multiplication rules.

As the variety of data types becomes richer, the task of keeping each data-processing function abreast of a complex program's latest changes becomes more problematic. Instead of attempting to modify every operation with which a new object might come into contact, the object-oriented programming approach merely attaches all rules for special handling to the object itself. That way each operation can't help knowing what to do whenever it encounters a new situation.

If I were to ask my secretary to "ship *this* to Sandy's office," her action would depend upon the nature of the object my "this" referred to. She might mail it, fax it, call for porters, or even carry it herself, depending on its weight, urgency, fragility, or value. Instead of attempting an exhaustive shipping discussion as part of organizing our office, however, we depend upon associations between object and operation—from heavy lifting for a desk to faxing or mailing for a one-page letter.

In the same way, while the code for an object-oriented program might still read:

```
Let a = 5
Let b = 6
Let c = a x b
```

the underlying operations would differ markedly from the earlier case. In an object-oriented execution of these instructions, the program would send a message to the *objects* located in registers a and b, *telling* them to *multiply themselves* together and ship the result to register c.

If the object in "a" were a vector, it would embody the rules for multiplying itself, both with vectors (as in the torque example), as well as with scalars (as when a ship reduces engine power), but would limit addition and subtraction operations to other vectors.

While the need to carry all this extra stuff around with each object consumes processing power, such an investment pays off in reliability and flexibility. In the latter arena, adding a new kind of medical reimbursement no longer requires modifications to the billing programs that must execute it. Say a company encourages employees to get checkups more often by footing the bill. An object-oriented program might still tell all electronic registration forms to "specify payment," but an employee's *sponsored registration* would "know" enough to arrange for payment by the sponsor rather than by the patient or insurance company.

Note how this object-oriented approach shields the billing system from the extra complexity that yet another form of payment might otherwise entail. Since whatever procedures sponsored bills require remain inside that *object,* we don't risk messing up billing on other services—as we might if a single unified program handled the entire job.

Today, modern software advances—such as middleware and object-oriented programming—allow the assembly of integrated business systems out of previously separated components. But far more lies ahead. As we look to the future, we can expect to see far greater levels of systems integration throughout our society—including within the daily lives of individual consumers.

Help in the House

For starters, I expect a growing number of home appliances to include a data port within a few years. Just like today's familiar electric power cord, each VCR, TV, telephone, air conditioner, and security alarm will plug into a common set of outlets so as to exchange information over the *home network* I mentioned in Chapter 6. Over time, smaller items—such as light switches, thermostats, and coffeemakers—should connect up as well. That way, you can turn them on and off when you're not there. Moreover, nonwired "appliances," like the computer in the family car, can join that same network via radio links.

Seeing a car maintenance special advertised on television, Joe Couchpotato could check the miles since he last had the brakes looked at—without budging from the popcorn bowl. Sounds farfetched? Only if we're willing to settle for a stand-alone system, meaning one that handles that particular task and connects to nothing else. The art lies in providing this service along with many hundreds of others. After all, if our TV-watching car owner had to remember how to access his vehicle's onboard data-storage system, he might just as well go rummaging through a drawerful of old repair bills.

Here, then, lies the systems-integration challenge. Suppose our car owner's "personal computer" normally handled such a request—made verbally, in this case, via the TV's voice-operated remote control. From its point of origin to the PC, from there to the telephone via the home network, then to the car's cellular phone via the public network, and finally to the car's computer via the car's own network. Small wonder that industry analysts see a vital role for middleware to help manage all this complexity.

As a simplifying layer, middleware acts much like a computer operating system does at a somewhat lower level. As you select a menu topic displayed in one of today's PC windows, you needn't concern yourself with where the file in question was stored or the data path that might bring it into view. The PC's operating system shields you from worrying about such details.

In the same way, middleware shields users from details at a more global level—notably the intricacies of individual machines, databases, and the multiple networks that interconnect them. Users need only concern themselves with the set of applications that best suit their individual needs. In particular, I speculate that the most useful of such applications programs will act as *intelligent agents* on their clients' behalf.

Intelligent agents take action on the basis of inferences drawn from their users' behavior. In one recent example, a popular word-processing program produces a full return address as soon as the user types its first few characters in the upper right-hand corner of a fresh page.

Looking toward the future, I expect a considerable broadening of the behaviors these intelligent agents can monitor. With data from environmental sensors already available for home security purposes, it takes no great stretch of the imagination to envisage applications that can distinguish between welcome and unwelcome guests.

I'd like to use a name other than "personal computer" for the handy home helper that I see running a typical twenty-first-century household. After all, nearly everything will *be* a computer in those days. Let's go through a day in the life of a twenty-first-century homeowner we'll call Michael and the electronic *home assistant* he communicates with for personal services.

Since such home assistants respond verbally most of the time, Michael began their relationship by selecting the voice in which the "home helper" would address him. Preferring British reserve with a dash of warmth, he settled on Richard Burton as a vocal model. In addition to holding up "his" end of conversations with Michael, "Burton" plays a dual role as personal secretary and full-charge housekeeper.

In the former role, Burton tracks everyone's whereabouts and activities (unless instructed to butt out temporarily). "He" regularly accesses the database monitoring the status of the house's inhabitants. That database, in turn, receives input from a sensor processor that analyzes audio, acoustic radar, and video data received from the sensors located around each room.

Acting as an intelligent alarm clock, Burton recognizes when Michael goes to sleep or wakes up by analyzing sensor data concerning his breathing patterns and movements in bed. Once awake, Michael can ask Burton most of the questions one might address to a faithful servant, such as "What do you think I should wear today?"

In addition to sending clothing to the cleaners via a robotic delivery service, Burton supervises the household robots that wash the laundry, hang clothing in the closets, and keep Michael's dresser drawers in good order. The information gained in that oversight role—together with weather reports, Michael's schedule for the day, the outfits he likes best, and a complete record of the ensembles Michael has worn in the past—combines to shape Burton's reply. If it mattered, Michael could probe further and inquire as to which suits each of the people he's likely to see that day have already seen him wear, and when.

In his housekeeping role, Burton makes sure that Michael remains pleased with the meals his kitchen prepares and serves—as well as seeing to the special needs of any guests. Even the best of machines breaks down, of course, so Burton also arranges for appliance repairs and preventive maintenance. In dealing with the vendors who supply household goods and services, Burton verifies the quality of each transaction and arranges for payment—keeping Michael informed as appropriate. If Burton himself malfunctions, a backup program on another machine in the network stands ready to take over.

As a secretary, Burton fields much of the hassle that would otherwise befall an unassisted human living in a complex world. Monitoring every receipt, warranty card, and bill, Burton maintains a complete on-line record of such items. When needed, this data can provide instant inputs to tax programs, medical insurance claim forms, and the like. So Michael can get an immediate answer to a question like "Where did I put the receipt for the hot water heater?"—no matter how many years have elapsed since its purchase.

But rather than proceed further with this fantasy, let's consider its reality. Each of the individual items in the scenario—sensors, actuators, and the networked computing needed to support their performance—seems likely in the early decades of the twenty-first century. A software layer, like the middleware programs we touched upon earlier, should allow the voice-controlled home assistant to link up with the wide variety of appliances, databases, and other artifacts that will support a twenty-first-century household. All these functions in an easy-to-run integrated package should make an attractive product. With "Burton" and his fellow home assistants available, I can't imagine many people asking "Who needs all that technology?" Unless of course, they enjoy doing their own housework.

Do You Know My Name?

Moving outside the home, I picture an environment resembling that of a small village served by individual proprietors and public servants who know each of its citizens by name. Of course, village life doesn't suit everyone. Depending upon personal preferences and circumstances, most of us would likely opt for a balance between convenience and privacy. So if privacy concerns—such as the thought of "everyone" knowing that you're renting an XX-rated video—fill you with indignation, please rest assured that the systems I'll describe lend themselves to convenient privacy-preserving options. Without sacrificing all their privacy, future inhabitants of an integrated world can enjoy the individually tailored benefits that small villages have traditionally offered their inhabitants.

Whatever its other advantages, the impersonal character of the modern urban environment makes few allowances for special needs. Coming to the United States when my brother and I were young children, for instance, my parents had a difficult time adapting to our English-speaking surroundings. Unable to read signs or ask directions, even the simplest errand confronted them with major obstacles. As today's electronic interfaces proliferate, however, technology can replace earlier barriers with bridges. In that regard, consider the multilingual character of Israel's automatic teller machines. When a recent immigrant from the former Soviet Union inserts her card, Cyrillic instructions appear on the screen in place of the normal Hebrew. The ATM card, in this case, identifies its owner's language preference—not just her account number.

But that's just the start. As founder of the world's most successful PC software company, Microsoft's Bill Gates may be excused for thinking of tomorrow's credit card as nothing other than a wallet PC. "When you want to board your airplane," he told a *New York Times* interviewer, "you sort of just walk by and it knows who you are." In other words, the device identifies itself to nearby servers, orchestrates unobtrusive verification of its owner's identity, and engages in transactions on that owner's behalf via a short-range wireless link.

For those who balk at carrying even a wallet-size PC around with them, consider the recent emergence of electronic smart cards I mentioned earlier. While they look just like the ordinary charge cards in present-day wallets, some of these devices already contain a linked set of computer chips sandwiched inside them. Slipped into a "reader" that supplies a keyboard and display, the combination becomes a full-fledged wallet PC—allowing its owner to review and employ whatever information the card contains.

Since such a card will likely embody enough memory to fill a modest-sized notebook, it can clearly accommodate all the functions of a walletful of today's plastic—and more. Still, some data would better reside elsewhere. Take medical records, for example. Rather then attempt to store high-resolution images of all the chest X rays you've ever had, the card would merely note their various locations—together with the means by which a physician treating you could obtain electronic copies.

And what about reliability? With networked computers and distributed files, vital information can't get lost or stolen. If one system breaks down, another will take over, using backup data kept automatically up-to-date for just such an emergency. As for theft, the smart card system helps its users identify themselves unobtrusively—via an owner's voice or fingerprint, for instance. Finally if the card itself gets lost, the owner merely engages in a somewhat more lengthy identification procedure over a telephone line connected to the physician's computer, among others. Nothing is perfect, of course, but such systems sure beat storing, shipping, and safeguarding paper records.

While such pocket-sized devices don't preclude a centralized gathering of individual medical records in massive databases, they work just as smoothly with more distributed systems. Individual hospitals could just keep their medical records on network-accessible database systems, then supply information to authorized inquirers. The required authorization codes would reside inside each patient's wallet, along with basic medical data, insurance coverage, and the like. With the insurance company contacted for reimbursement on the spot, the smart card's owner could pay any outstanding balance by shifting it into its bank-card role—filing away an electronic receipt for tax purposes as well.

Eyes and Ears

As systems such as these become more adept at serving users, entirely new applications should emerge. Ever since the days of Dick Tracy's wrist radio, wearing one's telephone has become a commonplace amenity in most futuristic fantasies. With just a little more technology than already exists, however, I speculate that reality can outstrip fiction—adding some features that the cartoonists have overlooked.

Think about having to decide which microphones to wear when packing for a twenty-first-century convention. I expect microphones to play a role as personal jewelry—much as wristwatches do today. Instead of a single microphone-containing lapel pin, belt buckle, broach, or tie clasp, I speculate that many people will opt for several microphones, located at various places on the body. In that way, the combination might phase its electrical outputs with respect to one another so as to aim the resulting acoustic array at an interesting target—much as present-day electronically steered radar antennas scan the sky.

Since wiring all these gadgets together would make getting dressed really awkward, I would expect each microphone package to contain a tiny radio transmitter, linking it with a wallet-sized cellular phone carried in pocket or purse. Naturally, the convention-goer would also wear a radio-linked earphone as well, thereby establishing a two-way link to home base.

In conjunction with that audio apparatus, a pair of iris-sized video cameras perched on the outside corners of the wearer's eyeglasses would complete the ensemble. Like the microphones, the video cameras would embody radio transceivers, linking them to home base via the same wallet-sized relay phone. Thus, the home-based computer would "see" as well as "hear" people and events as its owner encountered them—and store that information so as to serve as a prompter at repeat meetings.

What a boost for "nerdy" techies who can't remember names and faces! Over time, one can expect each user's system to compile an extensive video library to parallel the wearer's accumulated experience. By then, our users won't need to trust to memory alone. With a bit of prompting, each of them can rely upon a rich Rolodex-like address list.

For example, some might instruct their respective home-based prompters to pay particular attention to a face in focus when they say "It's really nice to see you." In response, the prompting computer would attempt to link the face in question to names it had overheard in prior conversations while simultaneously seeking verification from other sources—such as stored speech samples from a "voice-print" database. Much better than trying to steal a glance at a fellow attendee's name tag.

Through links to its owner's "secretary," which might be nothing more than another program running on the same machine, the prompter would peruse guest lists, appointment calendars, and various directories to anticipate the identities of likely encounters. After all, human faces don't look all that different from one another in a computer's "eyes." That's why extra information can help.

With aid from all this technology available for the asking, one could merely speak a second code phrase to receive the name that most probably goes with the appropriate face. In doubtful cases, the prompter might also supply the occasion at which its owner last met the suggested candidate. That would allow for a more graceful exit from any misidentification—as in, "Say, didn't I see you at the Giants game last week?"

On a more utilitarian note, students could employ similar systems as aids in retaining material from classes, seminars, and even informal discussions. With many of today's students already taking lecture notes on portable PCs, adding video clips hardly seems farfetched. Moreover, with all the means by which twenty-first-century citizens might access *information*, we can expect a shift in emphasis toward access to *expertise*. Helping people share knowledge with one another seems a worthy investment, especially if it also lets us have a little more fun at conventions.

"On the Road"

Just as systems integration can help improve life on an individual level, the same principles apply at the other end of the scale—the physical infrastructure upon which the totality of human society depends. Imagine, for instance, that every mass transit plan now on the drawing boards became a reality overnight. Would that fix all our traffic problems? I don't believe it would. After all, housing development across the United States and in most of the rest of the industrialized world has structured itself around automobiles for longer than half a century. Can you imagine a new office building, factory, school, church, or shopping center without its own parking lot and ready access to some highway? Even commuters who work in cities served by mass transit generally begin each day with a car ride to the local station.

Under such circumstances, can we blame Americans for their so-called love affairs with their automobiles? After all, no one looks forward to a trek to the bus stop in the rain. So cars keep multiplying faster than roads and parking spaces. In my home state of New Jersey, we added some three hundred thousand cars in the 1980s alone but made no appreciable changes in our road network. No wonder traffic keeps getting worse.

But I believe that we can apply systems integration to transportation. In that way, "getting to work" can mean more than just a choice between an electronic commute from home and a pair of hourlong struggles in bumper-to-bumper traffic. With a little bit of luck, proper planning, and the right technology, I think we can manage to finesse present-day highway congestion—and not by paving every surface with concrete.

In essence, my scenario calls for luring commuters away from their cars by offering them service that approximates that of today's chauffeured limousines. There's a catch, of course. Each of my limos must serve multiple passengers in order to make the system affordable. But technology can ease the burden considerably. Let's see how it might work.

Say that you're the customer. You call the dispatching service and give your name, destination, and some form of electronic credit. The dispatcher responds with a choice of times, usually separated by no more than a few minutes. You pick the one you like best—making sure to leave enough time to get your things together, because the limo arrives on the dot, and making it wait will cost you a stiff surcharge.

The limo arrives on the button—a plush van with comfortable reclining seats. The driver swipes your credit card through the reader as you board and take your seat. As the van picks up speed, you key in a request for your favorite beverage and settle back with a newspaper or chat with a fellow passenger.

The driver makes occasional stops to pick up and drop off passengers as you await your first transfer. The panel in front of you indicates the probable duration of your trip as well as the time you'll have to transfer. As the ride proceeds, an alerting tone informs you of a change. Instead of two transfers, you'll now be making only one, and you'll save a few minutes in the bargain. The routing system just found a faster way of getting you to work this morning.

As your van nears the rendezvous point, you can see that your next ride has just pulled up to the specially marked curb. A couple of its passengers are moving toward a covered bench to await a van coming up behind you. The entire transfer takes no longer than it does for a long traffic light to change, and off you go again.

Clearly, the success of such a system hinges upon its ability to transfer passengers smoothly. After all, few people would reject direct door-to-door service on demand at bargain-basement prices. How, then, can the "dispatcher" (human or computer) orchestrate this seamless rendezvous? Only through instantaneous and complete knowledge of every vehicle's location, the individual destination of all passengers, and traffic conditions—together with all the internals of the business itself.

In principle, at least, the components for such a system appear almost within reach. Satellite-based locators can continuously pinpoint each vehicle within a dozen meters, anywhere on earth. While they still carry hefty price tags, thousands of autos already employ them in Japan. With the continued decline in consumer electronic prices, we can expect them to cost no more than high-end compact disc players in a few years—and to achieve a similar level of acceptance from new-car buyers.

In the present mode of operation, each vehicle's onboard navigator brings up the appropriate electrically stored street map and plots the region of interest on a dashboard display. After the driver selects a destination, the system plots a route and prompts the driver with computer-generated speech. "You are approaching a motion picture theater on the right. Please prepare to make a left turn at the next intersection." If the driver goes the wrong way, the system waits for a moment (so as not to rattle that person in the middle of a turn) and tactfully adjusts its instructions to return the car to the correct route.

With this enabling technology as a starting point, it's easy to see how future taxi fleets can improve their operations. Just think how frequently today's dispatchers use the phrase "Where are you now?" over those scratchy two-way radios. Onboard navigators can make a difference. No more getting lost, and less time wasted stuck in traffic. By noting the progress that individual cabs report along various roads, dispatchers can spot traffic jams and redirect their colleagues around them.

In the van-dispatching application, each vehicle's locator would report its position on a regular basis to a central record-keeping computer. This record keeper could provide data to human dispatchers or, over time, evolve into a system that could take over much of the dispatching work itself. While humans have little trouble matching potential riders with available cabs on a one-for-one basis, the multipassenger, multivan scenario I depict would call for the full-time attention of a good-sized supercomputer and some very sophisticated software.

But supercomputers will inevitably get cheaper, and software more powerful—just as the cost of highways will continue to climb ever higher. Thus, transparent sharing of information should spare future commuters from twice-daily encounters with the multilane parking lots many of us now face.

Tying It All Together

The systems we have just reviewed—home helper, smart-card-based identity, eyes and ears, and limo network—illustrate just a few examples from a much larger list. Indeed, each aspect of modern life bristles with systems-integration opportunities. Rather than belabor the point, let's shift gears and go to the next level. True harmony, after all, demands more than a list of integrated systems.

To illustrate, we'll consider some of the interactions a futuristic user of these systems might generate in the course of a typical day. Suppose, for instance, that Michael wants to invite friends to a last-minute dinner at his house, as follows:

Michael happens to bump into a couple of old friends. In town for a business meeting, they have just that evening free, so he invites them for dinner—together with a few locals they knew from the old days—so they can make a party of it. With much to arrange, Michael might include his home assistant in the conversation to help with logistics.

Tuned in, the home assistant could send electronic invitations to the people on the guest list the friends assembled in the course of their conversation, suggest menu choices based upon the invitees' preferences, and arrange to match preparation to the number of acceptances. Even with the best of helpers, however, some tasks will need Michael's personal attention, so his schedule may well need to accommodate extra items. Some—like taking his dog to the vet's, for instance—require his physical presence. So transportation becomes a feature as well.

In all this, the integrated suite of Harmony Era products and services ought to provide Michael with a team of helpers, waiting discreetly at his elbow. So far we have seen a social secretary, butler, and personal chauffeur at work, but we can easily add to the list. Yet, as I hope this discussion illustrates, harmony concerns itself more with the team's cooperative spirit than with the size of its roster.

Experience teaches us that a massive system designed to be all things to all people will fall of its own weight. So Harmony Era systems integration calls for an open-ended number and variety of helpers in our work as well as in our personal lives. Humans, after all, have better things to do than to run errands between incompatible systems.

From the vendor's viewpoint, much of future commerce will involve the care and feeding of integrated systems. As we saw earlier, systems-integration opportunities have already spawned multibillion-dollar enterprises dedicated to this task—such as Andersen Consulting, for example. At the same time, increasing numbers of information-products and -systems vendors now package their offerings into integrated solutions.

In the past, the most significant examples of large-scale systems-integration successes came from single vendors. IBM's Systems Network Architecture and the Bell System's end-to-end network provide notable examples. However, in the future a different environment seems more likely. While earlier integrators could afford to create their own piece parts and set their own standards so as to assure interoperability—as if they ran the only game in town—those days seem gone for good.

Today, high-quality alternatives abound. Given the dazzling—and frequently bewildering—array of options, we needn't wonder why systems-integration consultants find such ready markets for their services. Able to combine the best components available and tailor them to individual needs, these integrators provide a valuable service and can clearly look forward to continued growth.

At the same time, we will surely see integrated, multivendor offerings from equipment producers as well. Both approaches offer advantages. On the one hand, integrators without products of their own to sell might well be seen as more objective. On the other, a particular systems-integration challenge could spur a manufacturer to create a new product that otherwise might not exist. One side, we might say, offers a broader list of what already exists, the other could more easily include what might be possible.

In actual practice, real life rarely matches such cut-and-dried distinctions. Andersen Consulting, for instance, manufacturers proprietary software, thereby going beyond what the marketplace offers. At the same time, the business-unit structure employed by most large information companies makes multicompany partnering easier. Depending upon the circumstances, one company's computer division might include networking equipment from another company in preparing a large bid—in preference to similar components from within its corporate family.

As I see it, a marketplace that values products capable of cooperating with one another will foster increased cooperation among vendors as well. System integrators will surely compete with one another at one level, but they must also ensure the interoperability of their architectures with those of other suppliers. Few customers will appreciate getting painted into a corner. In that sense, this highly competitive market for compatible systems reminds me of the management training course I described earlier. Would-be executives *contended* for choice assignments by *cooperating* with one another, knowing that the brightest future lay in store for those who demonstrated the best ability to cooperate with their colleagues.

As with the organization man of an earlier time, competition for relative advantage now demands cooperation. In the Harmony Era, however, the coexistence of competition and cooperation applies to the global economy rather than to the mere internals of a single organization.

With technology components more harmoniously arranged with one another, we can proceed to the remaining part of the harmony triad—improved relations between technology and nature.

CHAPTER 9

ENVIRONMENT

Just as the continued existence of human life depends upon our planet's ability to sustain it, the products of industrialized civilization can hardly survive beyond the limits of nature's forbearance. So harmony between technology and nature demands thoughtful attention. However much we may focus on the role of information in our economy, the fact remains that physical well-being depends upon the physical aspects of commerce such as manufacture, energy, and transportation.

As communities struggle with garbage disposal, traffic congestion, and health risks, consumer resentment begins to build. But looked at another way, this negative reaction contains a positive message. Every negative demand equates to a potentially positive demand for its inverse. And demand, after all, sows the seeds of business opportunities. Pressed hard by economic issues, most businesspeople, I venture to say, regard environmental concerns as more of a cost than an opportunity. And the scheme of things today still supports that belief to a large extent. But imagine what would happen if the government began to charge each manufacturer the cost of dealing with whatever garbage each automobile or bag of potato chips produced. Right now, someone else foots that bill, but times—as they say—are a-changin'.

It may take a long time to adjust the accounting system, but some jurisdictions—notably in Europe—have already begun to take the first steps in this direction. In any case, our economy as a whole must pay for taking out the trash and bear the cost of losing whatever natural resources our economic activities destroy. As regulation moves many of these costs closer to producers, manufacturing, handling, and reuse of materials generally improves. And as we'll see, the extra effort frequently results in net savings. Bridging the disconnects between technology and nature, therefore, seems to offer opportunities for doing well while doing good.

As the engine of life, nature uses air, water, and sunlight to rearrange matter—thereby sustaining the dazzling variety of living things that inhabit our biosphere. These creatures, in turn, further rearrange matter to help sustain themselves. A tree draws water and minerals from subsoil and carbon dioxide from the surrounding air and converts the mixture into new bark, leaves, and branches. A pair of robins builds a nest out of twigs, a chimpanzee strips the leaves from a twig to fish for ants, and a group of humans bore a hole from Folkstone to Calais.

Needless to say, we humans far surpass all other creatures in our ability to reshape matter to suit our purposes. The artifacts of our civilization abound everywhere—accompanied in many cases by unfortunate side effects. As we help ourselves to our planet's store of resources, we all too often dot our landscapes with gaping wounds and mounds of garbage.

As we all know, the human impact upon our physical environment has accelerated since the onset of industrialization. And modern technology places even greater leverage in our hands. While this extra power carries greater risk of negative environmental impact, it also offers us the means of redressing some of our past intrusions—and thereby forms the basis of new business opportunities.

For example, after rapacious fishing methods and the uncontrolled exploitation of rivers drove once-abundant Pacific salmon to the brink of extinction, a few commercial hatcheries tried to increase the presence of these beautiful creatures. Each year, they put thousands of tiny fish into local rivers and streams, hoping they would find their way to the ocean—and return as adults. When adult salmon did appear, however, each hatchery had only the most primitive ways of measuring the success of its efforts. Burdening a tiny fingerling with a cumbersome metal tag might well interfere with its survival.

More recently, however, a new tagging technology has largely solved that problem. Today, a growing population of salmon in northwestern rivers attests to the success of the restocking effort. And much of the thanks for that success goes to a neat bit of information ingenuity.

If you had happened to encounter a young atomic physicist named

Keith Jefferts in the early 1970s, you might have found him cutting notches into a pine dowel with a pocketknife. Jefferts's whittling had a serious purpose—large-scale models of tiny stainless-steel wire strands, each not much thicker than a human hair. Jefferts implanted this notching scheme on spools of the wire, thereby giving each spool a unique identifier. Most important, his code permitted the identification of a particular spool of such wire from an examination of a short (~3 millimeter) sample taken from anywhere along its length.

Today, salmon hatcheries buy spools of coded wire from Jefferts's company, Northwest Marine Technology, and—with the aid of a hypodermic needle—inject small pieces into their fingerlings' nose cartilages. Much easier than it sounds. Just grab a small fish, press its nose into a nose-shaped recess in the side of a toaster-sized box, and step on a foot pedal. That tags the fish. Through the use of multiple spools, each hatchery can implant identifiers that track variations in their processes—such as fish size and time or location of release.

To complete the process, the scheme employs magnetometers installed in most of the Pacific Northwest's salmon canneries. These instruments act very much like the "door frames" we walk through at an airport's security check. But instead of reacting to an airline passenger with a too large set of keys in his pocket, the cannery's equivalent beeps when a freshly caught salmon with a bit of wire in its nose passes through.

Alerted by the beep, a cannery worker merely slices off that salmon's nose, fills out a small form, and places both in a prepaid envelope. The envelopes make their way to a laboratory nestled in one of Puget Sound's prettiest islands. There, a technician retrieves the wire and decodes its identity with the aid of a microscope. As a result, the hatchery learns its alumnus's size and place of capture, while the cannery worker gets a modest monetary reward for making closure possible.

For each story of success in reversing the damage we've done to nature, however, many other needs remain unaddressed. Examples of reawakened interest in our environment notwithstanding, we need only look at the sorry state of Eastern Europe to see how much work lies ahead.

Despite obvious differences, the planned-economy bureaucrats who turned pleasant landscapes into smog-shrouded slag heaps share much in common with their opposite numbers in old-style capitalist enterprises. Just think back to depression-era photographs of Pittsburgh's smokestacks—or of coal miners in any one of a dozen countries. The difference, if any, lies more in degree than kind.

Throughout this book, we've touched upon the relationship between internally oriented paper shuffling and insulation from outside concerns. In that sense, the bureaucrats who created and operated Eastern Europe's industrial base according to a series of top-down five-year plans must surely rank among the most insulated of history's organization men.

But the spectacular failure of state planning in this instance should hardly cause us to dismiss the notion of planning itself. Like the bureaucratic insulation that allowed planned-economy managers to answer to no one but themselves, the single-minded pursuit of profit to the exclusion of all else has inflicted much the same damage throughout history. Great Britain had a sorry record in "dark and satanic mills" during the laissez-faire era that prevailed in the nineteenth century. That record should convince any thoughtful observer that environmental evils can also emerge from the opposite end of the economic spectrum.

While progress remains uneven, we see growing environmental awareness in the large majority of industrially developed nations. Need and opportunity go hand in hand. Consumers will demand—and businesses must provide—offerings that maintain our standard of living while respecting the physical environment that makes living itself possible. Since environmental shortfalls exist in all the principal areas of human activity—enough to fill a shelfful of books—the following sections touch upon illustrative examples rather than provide an exhaustive list. Each treats an important segment of our economy and ecology: manufacturing, recycling, nuclear energy, and transportation, respectively—examining conflicts in these areas and exploring prospects for remediation. To the casual observer, today's environmental dilemmas probably seem nothing more than expensive quagmires. But think back to an earlier situation. How many "casual observers" saw business opportunity in mail delivery before the creation of Federal Express? As we look at unmet environmental needs, therefore, some will surely see profit where others perceive only problems.

Green Manufacturing

While even the most benign appearing of manufacturing processes deserves careful scrutiny, there is one industry whose problems rarely make headlines. Most people regard it in a largely positive light. But let's take a look at some facts.

While safer substitutes have improved matters in recent years, this industry still uses enough poison each year to kill every person on the planet several times over. Little, if any, of this toxic stockpile gets recycled. And little finds its way to toxic-waste facilities. Where does it go instead? Aside from the small portion left as residue on its final products, the vast majority merely washes into the nearest waterway. The impact looms large. According to a 1994 EPA report, a whopping *72 percent* of all pollution found in "impaired" U.S. rivers stems from these effluents.

These same chemicals also exact a toll on the workforce engaged in this activity. Seemingly healthy—largely because they smoke less and exercise more than the population as a whole—these workers exhibit significantly higher rates of brain cancer, multiple myeloma, non-Hodgkins lymphoma, and a number of other cancers. Some studies show melanoma rates as much as six times those found in the public at large. While this particularly lethal form of skin cancer correlates with prolonged exposure to sunlight—another hazard associated with this particular occupation—various chemicals also act as causal agents.

An industry so small that it got overlooked? Hardly. It employs some two million workers in the United States alone and generates well over $100 billion of annual output. What then? Many readers will have identified this industry as the manufacturer of food and fiber we call *agriculture,* and the poisons in question *pesticides* and *herbicides*—substances designed to interfere with the (respective) life-sustaining processes of animals and plants.

While the technology associated with purposeful food production dates back to ancient Mesopotamia, the use of chemicals to control insects came into widespread use only in the twentieth century. Until recently, moreover, the methods remained quite primitive. "Lead arsenate kills insects," my (now-grown) children's *World Book Encyclopedia* tells us. Indeed it does. Since it contains both lead and arsenic, this once widely used chemical compound packs a lethal one-two punch.

According to a 1991 announcement by the U.S. Centers for Disease Control, the equivalent of a mere pinhead-sized lump of lead dissolved in a human bloodstream can produce a detectable effect upon the central nervous system. Larger amounts will kill you. Despite the fact that lead's far-reaching toxicity upon those engaged in its use was known in ancient

Greece, however, little of the search for acceptable substitutes predates the onset of environmental awareness of the 1970s. Until then, the encyclopedia note I cited above represented reality.

More recently, we've seen much improvement. Instead of pesticides that owe their toxicity to poisonous metals, most farmers now use chemical compounds that break down into nontoxic fragments over time. A big improvement, especially when we consider that this residue eventually finds its way into our food and water supplies. But while pesticides no longer constitute a major source of heavy-metal contamination, other sources also require attention.

Like most people, I still refer to the "lead" in a pencil, even though that lethal substance disappeared from these familiar writing instruments over a century ago. Since then, pencils have contained nothing but a mixture of clay and graphite. So for those of you who occasionally suck on the tip of a pencil for inspiration, please rest assured that you run no risk of lead poisoning.

Most of our commonly used batteries, on the other hand, contain large amounts of heavy metals. And battery use appears on the rise. If I were to convert my home to solar-powered electricity someday, I would need to store at least a three-day supply of electricity to protect my family against a spell of foul weather. Since we consume about twenty kilowatt hours on a typical day, and a car battery can store just about one kilowatt-hour's worth of electrical power, I'd need the equivalent of at least sixty automobile batteries in my solar energy system. But batteries only last a few years, on the average. Roughly speaking, therefore, my solar energy system could cause me to buy (and dispose of) one battery every few weeks. Even with careful recycling, I expect solar electricity to spawn a large increase in battery manufacture and the industries that support it unless we can find other means of storing electrical energy—such as flywheels, for instance.

Just as with our agriculture example, we must evaluate the full-stream impact of each production process in order to make the most environmentally benign choices available—no matter how much opinion-leader esteem that process might enjoy at the moment. While few people associate photovoltaic electricity production with heavy metals such as lead or cadmium, any system that depends upon batteries might well spur the entry of additional amounts of these substances into our biosphere.

Naturally, we can minimize the impact of any increase in battery use by incorporating recycling into this product at the outset. Moreover, when manufacturers incorporate recycling into their production methods, raw material savings frequently reward good behavior with a tidy profit.

Recycling

The depression year of 1932 seemed a particularly poor time to begin a new business, but at least one hopeful entrepreneur overcame the odds. In that year, Nicholas Marcalus—a thirty-nine-year-old self-taught engineer—established the Marcalus Manufacturing Company, a fledgling producer of food wrap, with a tiny amount of capital, a lot of hard work, and a remarkably simple idea.

Until that time, a housewife would wrap leftover food in one or more sheets of waxed paper. But the sheets stuck together, especially in hot weather. Annoyed at the trouble it took to peel a fresh sheet from the stack every time she needed one, Mildred Marcalus asked her inventor-husband to come up with a better idea. Nicholas accepted his wife's challenge and, in the process, earned the patent that served as the cornerstone of the business that ultimately grew into today's Marcal Paper Company—*rolled waxed paper in a tear-off dispenser.*

While he remained an inventor all his life, Nicholas Marcalus owed much of his success to his keen business sense. Branching out from food wrap into other paper goods, he bought up small paper mills to become less dependent upon suppliers than his competitors were. He consolidated his operations on a tract of land near Paterson, New Jersey. From there he found a neat way of cutting his raw material costs—by offering to dispose of the "waste paper" from the press overruns of local printing companies.

As sales volume increased, the Marcal company next approached municipalities for their paper as well, and it now receives over five hundred tons of high-quality scrap paper every working day from communities in half a dozen states. Remarkably, Marcal pays nothing for this material. The municipalities even truck it to the plant at their own expense. But consider the alternatives. Dumping ten tons of waste paper would cost the taxpayers of Morris County, New Jersey, about $1,500, while trucking it over thirty miles of highway to Marcal's plant costs little more than one-tenth of that amount. While more distant communities enjoy less of a saving, municipal authorities as far away as Pennsylvania and New England still realize a net profit from the same arrangement—especially as landfills become scarcer.

In order to fuel additional growth for its recycled paper products, Marcal now looks to the U.S. Postal Service's stream of undeliverable third-class mail as its next "virtual forest." In this highly competitive industry, after all, no one can afford to stand still. Thanks to innovative processes, the Marcal plants produce paper towels, napkins, and bath tissue from a blend of magazines, junk mail, office paper, and the like—without use of the chlorine bleaches commonly employed by others.

Water—a big item in the paper industry—gets purified and recycled. An on-site cogeneration plant produces heat and electricity, with the surplus sold to a local utility. All in all, a model of environmentally responsible behavior. But while Marcal's business style has earned it accolades from both state and federal EPAs, as well as from private environmental groups, the company's main source of motivation lies elsewhere. Listen to Nicholas's grandson, Peter, now vice-president of corporate communications:

> I think we all inherited the hand-me-down mentality of the
> Depression Era in which my grandfather lived and did business.
> Products that have a second life go along with the thinking of not
> wasting anything.

As environmental awareness grows, so-called hand-me-downs no longer carry the stigma of second-best offerings. Nowadays, in fact, one finds them in a growing number of unexpected places—including luxury automobiles.

"Put a bumper in your trunk," a BMW advertisement might say. That company's new 3-Series models now employ luggage-compartment liners made from ground-up bumpers as part of its recycling program. While the metal portion of a junked car normally finds its way to a smelter sooner or later, the rest usually ends up in a landfill. With its new Z1 roadster model, however, BMW introduced a car that *Business Week* recognized as "the first car ever designed properly from the outset for optimum recycling."

In particular, BMW engineers make extensive use of so-called *thermoplastic*—rather than *thermosetting*—plastics for as many of the new model's components as possible. The difference? The former melts much like a metal, so recyclers can grind up old parts to cast new ones, while thermosetting plastics, such as Bakelite, undergo chemical change during the production process. So one can't melt this latter material down for reuse. For some applications, like the handle on my teakettle, resistance to melting offers a clear advantage. But that same property provides no value in the structural composites required in automobile assemblies.

To encourage recycling, BMW designers avoid embedding metal parts in plastics, while also taking care to make the parts easy to identify and disassemble. That way, future consumers can enjoy better-looking scenery as they drive their new-looking recycled cars past fewer junk-yards and landfills. After all, by the year 2000 Europe alone will need to find a place for twenty million scrapped cars annually. What better place for them than back on the road?

Today, a growing number of Europe's automobile producers no longer regard disposal as someone else's problem. Volkswagen of Germany now assures purchasers of its Golf models that the company will take them back at the end of their useful lives. Furthermore, press reports suggest that the German government will mandate such a policy for a wide range of products in the near future.

In some cases, however, prevention alone can't rescue our environment from the consequences of pollution. Among these cases, one stands far above the rest—and presents us with the most challenging recycling task in history.

Nuclear Power and Problems

If the neighbors complained when Pierre and Marie Curie took delivery of a ton of *pitchblende* in their turn-of-the-century Paris apartment, the media of that day paid no notable attention. Instead, their fellow Parisians probably shrugged off the unusual behavior of this most unusual young couple. Both physicists by training, the Curies had purchased a wagonload of ore so as to isolate and identify the factors that gave this substance the source of its radiant energy.

With every cupboard filled to overflowing, the Curies set about reducing the radioactive ore into its constituent parts. Years of backbreaking labor followed—along with discoveries that changed the course of human civilization. The discovery of radium—so named by them because of its radiant properties—ranks as their most celebrated achievement, but there were many other achievements as well.

Among these, the Curies' discovery of polonium—which they named after Marie's native country—strikes me as the most poignant. Years afterward, Marie Curie still recalled the naive joy they both felt when, happily exhausted, the two scientists gazed at the glowing residue of an arduous refining process. At the time, of course, neither realized the danger posed by the contents of the pan before them.

Polonium ranks as the most radioactive—and therefore the most lethal—of the naturally occurring elements. It outstrips even plutonium in this respect. Constant exposure to this and to the other materials in daily contact with her skin, on clothing, and scattered throughout the apartment could hardly have failed to contribute to the leukemia that ended Marie Curie's life when she was forty-seven. But before she died, she had the unique experience of witnessing her daughter's own award of a Nobel Prize.

Irène Joliot-Curie was born—and lived her early years—in her parents' apartment-laboratory. A one-year-old witness to the discovery of polonium and plutonium, she devoted her life to the study of *radioactivity,* a term coined by her parents to describe the phenomenon that earned them immortality and ultimately caused their deaths. Tragically, leukemia—the form of cancer most closely associated with full-body exposure to radiation—cut Irène's life short as well. She died in 1956 at the age of fifty-nine.

From a single home-laboratory to the vast enterprises of the Manhattan Project during the Second World War, experiments yielded new knowledge and new technology—most notably the atomic bomb. In the years that followed the bombing of Hiroshima and Nagasaki, even more destructive weapons appeared—so-called *hydrogen bombs,* because their energy came from the *fusion* of several lightweight nuclei into a heavier one rather than the *fission* of heavy nuclei such as uranium and plutonium. As the power of individual bombs grew, so did their number as well as the number of countries that possessed them.

But the powers unleashed by nuclear technology found other applications as well. These days, anyone interested enough to secure a few back issues of *Scientific American* can gain a pretty good idea of what it takes to make at least a rudimentary atomic bomb: distribute a bucket of plutonium around the inside of a hollow sphere made of TNT and cover the combination with a thick steel shell. Setting off the TNT will create a blast that drives the plutonium inward and holds it together long enough for the required runaway chain reaction. Without an implosion to hold it together, the chain reaction's heat buildup would force the plutonium apart, thereby stopping the reaction before it got to cataclysmic levels. If, on the other hand, one pumps a coolant through the *pile* of radioactive material, the heat can escape and the reaction proceeds until the fuel loses its power to create more energy. The heat, then, becomes available for other uses, such as propelling warships, generating electricity, and—more recently—powering spacecraft.

In the last years before her death, Irène Joliot-Curie witnessed the launching of the first nuclear reactor designed for civilian use as an electric generator. On Labor Day 1954, President Dwight D. Eisenhower broke ground for a sixty-thousand-kilowatt demonstration plant at Shippingport, Pennsylvania, to be constructed by Duquesne Light and Westinghouse, in partnership under the supervision of Admiral Hyman Rickover, father of the U.S. Navy's nuclear-powered warship program.

At the time, it must have seemed like a free lunch. Instead of turning a ship's propellers, the steam generated by the reactor would turn the shaft of an electric generator. The generator manufacturers had only to take the navy's blueprints and make everything a bit larger. But economics proved otherwise. With coal-fired plants producing electricity at give-away prices by today's standards, small-sized nuclear plants simply couldn't compete. Westinghouse and GE—its rival in this new technology—went back to the drawing boards.

In 1963, GE offered fixed-pricing nuclear plants capable of producing in excess of half a million kilowatts apiece. And Westinghouse soon followed. The utilities started buying, but the manufacturers' bottom lines suffered fearful consequences. Over the next three years, the two rivals produced something like a dozen plants and lost $1 billion in the process.

But neither gave up. Instead they scaled up the power to the *one-million-kilowatt* range and worked at reducing costs whenever possible. And so a technology designed for one environment got stretched into another. The navy required a compact design and cared little about the cost per unit. More important, they could count on a staff of superbly trained and highly disciplined operators, working under the most careful oversight imaginable. Quite a contrast to life in an average power plant.

Unfortunately, enthusiasm—rather than caution—carried the day. Sales of nuclear power generators surged, with a new plant ordered on the average of every two weeks. But corner-cutting began to show as well. By the mid-1970s, the average U.S. nuclear power plant experienced one unplanned shutdown every six weeks. By contrast, each of Japan's present-day nuclear plants runs almost twenty times longer between glitches—and they do so under more stringent criteria.

The Japanese, of course, enjoy two advantages—a fervent devotion to quality and access to the lessons taught by U.S. failures, notably the pileup of human errors at Three Mile Island. Until that accident woke them up, much of U.S. nuclear safety oversight focused on million-to-one scenarios—such as a direct hit by a crashing 747 or a Los Angeles–sized earthquake in regions with no history of seismic activity. Yet they all but ignored human-error situations, such as operators assuming that a flashing danger light indicated trouble in the *alarm system*.

While the U.S. nuclear power industry still lags far behind the reliability standard set by Japan, even its critics—such as the Union of Concerned Scientists—acknowledge significant progress. So much so, in fact, that the area of primary public concern has shifted to the disposition of spent reactor fuels. Accordingly, much of today's nuclear power debate centers around a plan calling for deep burial in an arid region near one of the U.S. government's former atomic test sites.

While what I've learned about the plan convinces me that present-day technology could assure burial with far less hazard to humans than naturally occurring deposits of radioactive ores, I nevertheless feel that a more attractive solution to this issue lies nearer at hand—recycling. As it undergoes nuclear reactions, the uranium in a typical fuel rod turns to plutonium—itself capable of producing more energy when properly packaged with other elements. Unfortunately, plutonium also provides the easiest route to a do-it-yourself A-bomb. That's why many people regard such recycling with suspicion. So much so that the Carter administration placed a flat ban upon this practice—at least as it applied to the U.S. power industry.

Paradoxically, countries with no other path to nuclear weapons—such as India, Israel, and North Korea—appear to have availed themselves of this means of entry into the nuclear club, while the U.S. owns a huge stockpile of weapons far more powerful than mere plutonium bombs. Would more plutonium in U.S. hands really change the world's balance of power? Instead of letting it cook itself out inside a mountain for the next hundred thousand years, why not reprocess spent U.S. fuel rods and put the resulting plutonium to productive use?

Irrespective of the merit of such arguments, I see a likely shift toward nuclear recycling for a different reason: the need to deal with the former Soviet Union's stockpile of nuclear weaponry. According to a Worldwatch estimate, dismantling those weapons will yield over one hundred tons of plutonium and similar bomb-grade nuclear materials. Some people propose that they be launched into deep space or dropped into the core of an active volcano, but the feasibility of such schemes remains untried. Transmutation inside a specially designed power reactor, on the other hand, exercises an already established disposal technology—destroying plutonium via nuclear reactions rather than creating it.

Depending upon the reactor design chosen, such transmutation can produce varying amounts of power. At one extreme, a recent Japanese design can consume well over one ton of plutonium during each year of operation, while producing roughly the same amount of electric power as an average power plant. Since Japan plans to construct some twenty new nuclear power plants in the next few years—raising the percentage of electricity generated by this means from 25 percent to about 50 percent—dedicating just a fraction of these new systems to rapid plutonium consumption would neutralize the entire Soviet stockpile in a few years.

Until recently, weapons-related plutonium exceeded the amount contained in spent fuel rods, but the combined effect of treaty-driven limits to the arms race and the steady pileup at U.S. atomic power plants has now reversed the situation. Today, spent fuel rods account for some three hundred tons of plutonium and the amount continues to grow. Must the U.S. suffer from this self-inflicted environmental wound? Despite Europe's lack of vast tracts of arid desert, it faces a far more easily solved long-term disposal problem. Germany, Switzerland, and their neighbors merely ship their spent fuel rods to France for recycling—thereby saving themselves from the "what-do-we-do-with-the-waste?" headache and reducing the amount of newly mined ore their power plants require.

The French, for their part, reclaim plutonium and similar material from the spent rods and pack it into fresh ones. In the process, a relatively small amount of extremely radioactive (and economically useless) material comes off as a liquid. Converted into glass pellets to simplify handling and storage, this material must be safeguarded for at least two hundred years—long enough for it to lose some 99 percent of its present level of radioactivity but only a minute fraction of the time required for a similar decay in plutonium.

With France already producing some 75 percent of its electricity in this manner and Japan 25 percent and climbing toward 50 percent, a sizable group of U.S. scientists and engineers still hope for an increase from this country's 20 percent. But without a resolution of present-day nuclear waste issues, an increase doesn't seem to stand a chance of public acceptance. With recycling, on the other hand, some environmentally thoughtful consensus might form around the Japanese model.

In addition to a nuclear power system that ought to make other countries' operations ashamed to call themselves high-tech, Japan's approach to energy includes commitment to renewable energy sources (such as a firm plan to put solar hot-water heaters in half their homes by the year 2010) and far less profligate use of energy. Average Japanese citizens consume electricity at barely one-third the rate of their American counterparts. A fine example of a harmonious approach, I'd say.

No one technology alone appears able to do the job of creating energy. While solar energy ought to play a far greater role in everyone's energy future than it does today, even this much-touted technology imposes significant environmental cost. According to Worldwatch Institute, enough solar power plants to meet U.S. energy needs would require some fifty-nine thousand square kilometers of land—almost three times the size of my home state of New Jersey—let alone issues of cost, energy storage, and technical feasibility.

While the nuclear-versus-solar debate may provide some airing of real issues, adherents on both sides seem all too ready to adopt the "my-technology-solves-the-problem" stance. While such an attitude can make one feel good, it also obscures needs for further improvement. And needs, after all, offer the surest pathways toward a healthier economy.

Today, activism continues to garner a disproportionate share of media attention. But less noted events have produced greater impact. In particular, the U.S. electric power industry can no longer count on exclusive franchises. As a result, many once captive utility customers can shop around for their electricity—generally by having it shipped in from out of state over existing power lines.

Like the deregulation of telecommunications that occurred a decade earlier, the introduction of competition in the power industry will surely spur innovation. At the time of this writing, for instance, the Holmdel, New Jersey, laboratory I worked in for many years was slated to get much of its heat and electricity from a highly efficient *fuel cell*, which creates electrical energy via a chemical reaction rather than combustion. But Bell

Labs researchers can't claim any of the credit. The local power company merely selected that facility for an early introduction of this promising new technology. Since fuel cells can deliver almost twice as much useful energy from each pound of fuel as do conventional power plants, our utility enhances its competitive position with respect to would-be intruders with surplus power to sell.

Much as entrepreneurial mini–steel mills have spurred overall improvement in our once complacent steel industry, the advent of competitive offerings in the power industry should lead to better and more benign production and use of energy. While change will bring some short-term dislocations, the long run looks positive. Customers will enjoy the lower prices that competition normally brings, and creative-minded vendors will find new opportunities. And we will all benefit from a reduction in the stress that energy production—in all its various forms—imposes upon our environment.

Transportation

While proposals for new power plants invariably draw protests, who would think of denying one's neighbor the right to buy a car? And with good reason. On the surface, at least, today's automobiles seem a triumph of technology: relatively inexpensive as manufactured goods go, far more dependable than their earlier counterparts, more diverse in design, and brimming with helpful new features from antilock brakes to recorded voices reminding drivers to take their keys. And, as our cars have become more dependable, so has our dependence upon them.

But look at what's happened. As late as 1959, social and recreational driving split the lion's share of family car use with commuting. The remaining category—shopping and personal business—placed a distant third. By 1990, however, that picture had reversed itself, thanks to a whopping 216 percent gain in shopping and personal business.

With individual members of suburban households increasingly obliged to pursue geographically dispersed errands, multiple vehicles have become an everyday necessity. In the past two decades, the percentage of U.S. families with three or more cars has quadrupled to one-fifth of all households—and even more in the housing belts ringing our major cities. What a waste. According to the U.S. General Accounting Office, productivity loss from highway congestion alone now costs the country an estimated $100 billion a year.

Given such enormous social and economic costs, we needn't wonder at the calls for action. Unfortunately, much of the resulting effort aims at piecemeal solutions, some even counterproductive. According to the Worldwatch Institute, the use of vegetable-produced gasoline substitutes—such as ethanol from corn—can actually increase air pollution when one takes the entire production cycle into account. But who could argue against the lighter-weight more fuel-efficient cars now on the drawing boards? Or the smart highways that promise to provide extra capacity over existing roads?

While Intelligent Vehicles Highway Systems (IVHS) promise to add more "lanes" without extra concrete, the resulting effects on land use may well mirror the dreary results of our earlier infatuation with freeways. Suppose we could double highway capacity? What about city streets? Just doubling the parking spaces and garages of a modest-sized downtown like that of Washington, D.C., would consume over three square miles and cost some $6 billion. Not to mention what it would do to the quality of life there.

I could go on, but the point seems clear. Merely improving existing vehicles or their highways may well make matters worse. On the other hand, a holistic approach to IVHS technology can help guide vehicle use along socially responsible paths. With automatic toll collections via smart cards, congestion-dependent prices could enable traffic managers to adjust fluctuations in vehicle flows to healthier levels—and encourage the use of mass transit. Moreover, advanced IVHS capabilities would allot real-time priority assignments to buses, car pools, and van pools over single-passenger cars, thereby boosting multipassenger ridership.

With more traffic concentrated in mass transit, the old saw about "location, location, and location" as the three most important ingredients in business success should encourage a reversal of today's institutional proclivity toward suburban sprawl. In the process, air pollution, noise, accidents, and other insults upon our environment should diminish as well. And as our transportation system moves into a more benign alignment with our environment, new business opportunities will flow from better uses of human and material resources.

The Bottom Line

While much of the business community still takes a negative view of environmental concerns, others see such dissatisfaction in a more positive light—as a source of clues to potential demand. Here, the trick lies in identifying profitable countermeasures. In the Cisco Systems story I cited earlier, for instance, dissatisfaction with an inability to communicate led to the router that launched that wildly successful business. On the environmental front, the above-cited $100 billion now lost each year in U.S. traffic jams ought to motivate vendors of more efficient alternatives—such as the computer-dispatched van system I described in Chapter 8.

In the past, most economic dissatisfaction stemmed from an inadequate supply of things. But mass manufacture has solved that problem so well that we worry more about overcapacity than scarcity. Now, when one nation accuses another of "dumping," the charge generally involves high-quality goods rather than noxious waste products. With traditionally oriented companies still locked in head-to-head competition in existing product areas, it is easy to overlook the desire for a better life that created the demand in the first place.

Today, most of the channels that serve our society's material needs seem filled to overflowing. Entrepreneurial attention, therefore, should shift toward less tangible aspects of this desire for a better life. As a result, we should see greater marketplace emphasis on environmental improvements—a key value source in the Harmony Era. Rather than an adversarial conflict between human handiwork on the one hand and nature on the other, Harmony Era business practice should help us to live without destroying that which supports life itself.

CHAPTER 10

ISSUES

In the Harmony Era, technology will work in greater unity with our environment, with itself, and—above all—with its human users. What then? What pitfalls lie ahead as we prepare to take advantage of new benefits? How do we prepare our businesses, as well as ourselves, for this emerging era? What must we do to shape this new era to create a better and more benign world for all of us?

While no one can predict the future with certainty, we can look to the past for useful lessons and to the present for suggestive trends. In that spirit, let's return to the quantity-quality and quality-harmony contrasts that I introduced in the first chapter of this book. As we've seen, key features of the Quantity and Quality Eras contrast as follows:

QUANTITY ERA	QUALITY ERA
Mechanical technology	Programmable control
Economies of scale	Economies of speed
Advance-planning emphasis	Customer feedback emphasis
Hierarchical organizations	Team-based organizations
Value from volume	Value from performance
Technology islands	Technology overlaps
Environmental exploitation	Environmental concern

In Chapter 2, we reviewed the transition from mechanical technology in manufacturing to programmable control. With the widespread use of computers as monitor-controllers, producers could spread the costs over multiple tasks—thereby negating the old regime's economies of scale. Today, manufacturers live or die by the speed of their operations rather than by their size. In the personal computer business, for instance, producers must now turn over their entire inventories every sixty days—just to keep their costs in line with industry norms.

Rather than the rigid advance planning required by special-purpose machinery—built from scratch to perform exactly one way—programmable machines needed contact with their neighbors on the production line so as to mesh their operations. And this coordination boosted quality—so much so that the products of old-style manufacture quickly became obsolete.

Not so in offices, however, where entrenched hierarchies continued to reign. After all, the leaks weren't in their end of the boat. While computer-supported factories outperformed their competitors, computers did little to simplify paperwork. And even if they had, the goals of office work remained firmly focused on internal goals. Merely speeding the process would have meant moving faster in the wrong direction. Small wonder that most Quality Era offices remained mired in paperwork, with or without computers.

But as we have seen, the growing merger of computers and communication gives direct information access to many people who once depended upon chains of human intermediaries. Thus, while to my knowledge no computer has yet managed to replicate the performance of a single office worker, the right combination of computing and communications can frequently replace entire departments.

As businesses take advantage of technology, lower their costs, and improve quality in head-to-head competition, markets become saturated and opportunity moves elsewhere. So this combination of continuing technological advance, a workforce in need of reemployment, and a surfeit of goods undermines the value system established in the Quality Era. And the signs of further change appear evident. In particular, as growing numbers of would-be billionaires find profit in underserved areas of emerging need, we need to revise our economy's value map.

Recall how U.S. TV manufacturers couldn't give away their sets at any price once the public learned of the superior performance (in terms of lower trouble rates) offered by Japanese competitors. With every TV set in the store virtually certain to work perfectly these days, however, performance alone no longer assures a profitable sale. And so, new attributes—notably the ability to link up with other appliances and home controllers—will become the hallmark of value in yet another era.

For those with an historical turn of mind, we might have constructed a similar pair of columns contrasting the earlier cottage industry era with its successor, the quantity era of mass manufacture—a transition that history refers to as the *Industrial* Revolution. Like blockbuster movies, however, revolutions lead to sequels. In common usage, the decades that marked the transition from quantity to quality generally rank as the *Information* Revolution. But that shift set the stage for further transition. Will the change brought about by direct information access prove profound enough to earn it a name of its own? With pundits playing fast and loose with "revolution" labels, I'd say that something like "the Second Information Revolution" seems appropriate. Let's look at the contrasts. Comparing quality with harmony, as before, we have:

QUALITY ERA	HARMONY ERA
Programmable control	Direct information access
Economies of speed	Economies of convenience
Customer feedback emphasis	Personalization emphasis
Team-based organizations	Architectural organizations
Value from performance	Value from coherence
Technology overlaps	Technology merger
Environmental concern	Environmental renewal

As we have seen, the transition to harmony shifts the locus of value from individual products to integrated service. To a large extent, each customer becomes a partner in value creation by participating in "design" at the point of sale through interaction with optional offers—as with the individually customized TV set I mentioned earlier.

In the final analysis, the move toward harmony brings greater coherence to the total value-creation process. Technology becomes more aligned with its users and with the surrounding environment; businesses and their employees shift their focus from internal matters to their customers; and individual citizens become better able to live and work as unique human beings—contributing to, and benefiting from, the economy of a new industrial era.

In this book, I have tried to sketch the outlines of the emerging

Harmony Era as I see them. But much remains open. Social and economic choices made by all of us will surely color the picture in significant ways. In that spirit, this final chapter touches upon four of these issues—two of which concern technology, and two its human users.

From past experience, we know that the more useful a technology becomes, the more use we will make of it. And the more we use something, the more we come to depend upon it. As we rely even more heavily upon an interconnected web of such artifacts, will increased convenience also bring increased risk? And so, prudence requires purchasers to ponder the question of *reliability* in making technology selections.

In addition to assuring themselves of a safe margin against possible technology failures, most inhabitants of tomorrow's networked world will likely wish to reduce their exposure to unwelcome intrusions upon their *privacy*. Each of the electronic transactions that pervades everyday life creates data—though few of us devote much attention to that fact. In the future, however, I expect to see more people availing themselves of the privacy-enhancing capabilities that emerging technology will offer.

As we secure our systems against failure and our activities against surveillance, I believe we ought also to secure our personal sense of self against intrusion from false stereotypes regarding technology. While most people view the *anthropomorphization* of computers as little more than harmless terminology, I hope we can avoid the *mechanomorphization* of humans. If we come to see computers as electronic brains, might we not then see our brains as mere biological computers? Personally, I'd welcome a healthier division of labor. By assigning our mechanical chores to our machines, we can focus attention on the aspects of life at which humans excel.

Finally, with tools of every kind enhancing the skills and productivity of their possessors, will further improvements in technology widen the *economic gaps* that already divide the "haves" from the "have-nots"? If we hope to create a stable world for our children, we must find a better mechanism than mere charity to counteract the spread of poverty. While far from easy, I do not see the problem as hopeless. Unlike physical possessions, after all, our store of knowledge and information needn't diminish when shared with others.

As we review these issues, let's remember that all technology flows from the work of human hands and the minds that direct them. So the shape of life and work in the coming decades remains up to us.

Reliability

With businesses hoping to sell a mass of complex technology to one another as well as to the consuming public, one simple question must come before everything else: "Will it really work?"

When it comes to new technology, I generally play the role of professional pessimist. "What could go wrong?" I ask myself. But take automatic teller machines. When I first heard of these, I reflected that human tellers sometimes make mistakes—or even on occasion run away with the depositors' cash. So a machine that could take my money and alter my account balance seemed scary. As a chronic worrier, I found the ATM's wide acceptance surprising. With all my worrying, however, I missed one possibility of ATM misuse—one that some technically adept crooks eventually thought of and exploited.

In planning to block ATM fraud, conventional wisdom (including mine) focused on an unauthorized person gaining access to an ATM. Nobody thought about an unauthorized ATM gaining access to a person. So, when a legitimate-looking ATM machine appeared at a Manchester, Connecticut, shopping mall one day, passersby in need of cash naturally attempted to use it. None succeeded, however. The machine rejected their requests and sent them elsewhere—but not before it had recorded the numbers on their cards as well as the owners' PIN numbers. Most users shrugged off the encounter and tried their luck on another ATM. No one paid much attention—until a flood of unexpected withdrawals started appearing on bank statements.

While the amount of money lost cannot compare with that of even the smallest S&L failure, the story made front-page news nationwide—testimony to the level of public trust in ATMs. A case of "man bites dog," I'd say. An unusual occurrence, but not one likely to be repeated. Alerted to the problem, ATM software engineers will simply add a feature that assures the user of the machine's legitimacy instead of just the other way around.

Even with hindsight, it seems mean-spirited to fault ATM system designers for this oversight. After all, ATMs seemed destined for service in the lobbies of well-established banks. Who foresaw their spread to supermarkets and street corners? But the design of truly reliable technology will require just such foresight. In order to earn marketplace acceptance, future systems should survive every possible use, and misuse, in order to conform to our harmony standard.

With the inevitable escalation of complexity that the future holds, some people would call a halt to progress if they could. But such a path entails unacceptable drawbacks. Just look at the fight against AIDS, for example. In probing the protein coating that shields that virus from attack by the human body's natural defenses, molecular biologists rely on systems that define the atom-by-atom shape of the coating's surface. New, more powerful, instruments enter the struggle almost every month. Would anyone wish to stop that flow? I know of no way to maintain the innovation engine for medicine alone and deny it to other needs—such as education, transportation, and industrial productivity.

Fortunately, computer science continues to provide a steady stream of advances—such as the object-oriented programming I mentioned earlier. These advances, in turn, allow software engineers to improve the quality of their output. Much like the near perfection we find in today's hardware, software benefits from step-by-step attention to detail, together with sophisticated testing strategies. Indeed, trouble rates have been so reduced that something on the order of only a single typo in a six-foot shelf of books seems quite feasible.

While one bug per million lines of text may seem wonderfully small, would-be vendors of integrated systems ought to look at how many programs such systems will interact with during the course of a day. Consider my experience in moving to a PC-based electronic mail system. Before then, my "mail" sat on a shared corporate computer that both my secretary and I accessed via terminals in our respective offices. Under the new system, on the other hand, the mail came directly to my PC, thereby making it more difficult for her to keep track of my electronic correspondence.

To address this problem, I requested a "simple" fix from my system's administrator: automatic copies of all my incoming and outbound messages. But look at what happened. As long as I addressed my messages to third parties, the system worked perfectly. But when I addressed a note to "Julia," she got the original as well as a copy. Bad enough. But then the system sent her *a copy of the copy* as well as a copy of *that* copy—and so on, until her mailbox exhausted its ability to store additional copies.

Once the problem appeared, I merely had to amend my request to read: automatic copies of all my incoming and outbound messages—*except those addressed to Julia*. That way, both our separately reliable mail programs could interact reliably with one another.

As our computers proliferate and interact, the widening range of tasks we assign them will spawn encounters with ever-increasing amounts of software. With all the diligence in the world, how can we believe that the team inspecting a piece of software found every last bug? Better yet, how can we determine how many bugs they didn't find?

Surprising as it may seem, statisticians have devised a straightforward way of answering the latter question. Instead of assigning a single inspection team to examine the quality of some new piece of software, modern testing methods suggest giving that same assignment to two independent groups and then comparing the results. Suppose the first group found 110 bugs and the second 111. What would that tell us about the actual number of bugs?

To answer that question, we would need one more piece of information. How many of the bugs that the first team found also appeared on the second team's tally? If the first and second teams had 109 identifications in common, we could safely assume that we had missed only a few—and even estimate the likelihood of finding the rest with additional searches.

But suppose the two teams had just one single identification in common. That would imply a much larger number of uncounted bugs—something like ten thousand. In that case, additional testing would do little good. Better first to improve the process that produced the code. While perfection lies beyond our reach, we can nonetheless devise strategies that anticipate certain levels of reliability and then guard ourselves against the consequences of failure—by such measures as providing backup systems, for example.

To avoid potential miscues in the first place, software architects must not only assure the internal consistency of their offerings but also engage potential users in as much interaction with prototypes as possible—thereby enhancing user participation in the creation process itself. But even the most far-sighted of designers and the most articulate of users can't go beyond whatever knowledge exists at the moment of creation. The best of systems, therefore, still require human pilots to navigate uncharted waters.

No matter how reliably a system works in its predetermined context, no amount of forethought can exhaust all other possibilities—such as the one we saw in the ATM fraud example. Unlike computers, humans rarely address problems by applying a rigid set of rules. More likely to make small mistakes than computers do, we have far less trouble grasping the larger picture. And this difference suggests an important division of labor. Machines handle the nitty-gritty details, always under guidance of the broader perspective provided by human intelligence.

To ensure reliability, then, ease of use must go hand in hand with systems integration. As links between systems boost their complexity, the human ability to supervise them must keep pace. Ease of use thus becomes more a matter of necessity than mere convenience. A highly computerized world will require the exercise of human judgment. As a result, many of the people now employed in paperwork will find new employment opportunities.

Imagine that you had a choice between two automated transaction systems, only one of which offered the services of a helpful human in case of difficulty. Which one would get your business? While some people might choose a fully automatic system for incidental purchases, I can't imagine entrusting my savings—let alone my safety—to a system beyond all human control.

In that connection, our ease-of-use concept must embody ease of control. That's one reason why my next-century computerized transportation system includes the use of human drivers. With helpful, empowered humans available throughout tomorrow's integrated environment, one need never be without the services of a broad array of computerized helpers—provided, of course, that one welcomes such a high level of attention.

Privacy

For a variety of reasons ranging from personal choice to economic advantage, individuals and businesses may wish to limit access to information concerning their actions and circumstances—hopefully without significant cost or loss of convenience. After all, with so much extra computing power available from emerging technology, shouldn't systems providers allocate some to shielding their customers from unwelcome prying?

Unless consumers opt for active countermeasures in large numbers, some loss of anonymity seems inevitable, if for no other reason than record-keeping improvements. For example, I can't imagine that many would-be renters of videotapes would opt for leaving a cash deposit rather than revealing their identity to the store clerk. And millions of such everyday choices build databases. As the role of personalizing services grows, so will person-specific records.

In the absence of privacy concerns, I could imagine everyday life in the small village environment I mentioned earlier—one in which each citizen's actions take place within the sight of fellow "villagers." While automatic credit and customized services offer undeniable advantages, however, life in such an environment won't suit everyone. That's why inhabitants of real villages in the past frequently removed themselves from snoopy neighbors. But how does one move out of a networked village that girdles the globe?

On the one hand, the Harmony Era version of village life offers convenience of location-independent access to friends, colleagues, and resources. But not everyone welcomes full-time attention from a system that tracks individuals and their actions. Fortunately, modern encryption technology should allow future "villagers" to select whatever mix of privacy and access best suits them. While much of this technology evolved to fulfill military requirements, wider usage seems likely—provided that large-scale demand develops.

But do most of us really value our privacy enough to spend money? When the advent of CD-ROM technology made it possible to encapsulate encyclopedia-sized compendia on a small disk of plastic, Lotus Development attempted to introduce a nationwide consumer directory for PC users: names and addresses of tens of millions of U.S. households together with pertinent marketing data. A huge outcry ensued, and Lotus quickly withdrew the product. But no one seemed to care that the same information has long been available to anyone willing to pay a modest fee.

Privacy, it seems, remains unnecessary—as long as consumers can pretend it exists. Perhaps, but I have a sneaking suspicion that some entrepreneur will devise a way of convincing customers to buy privacy protection—something that they didn't realize they needed. And when that happens, other vendors may well kick themselves for not having made more aggressive use of encryption technology in their own products and services.

To talk about encryption, let's turn the clock back about fifty years. It's 1943. I'm ten years old and wearing a badge that identifies me as a member of Captain Midnight's Secret Squadron ("Ten cents in coin or stamps, and two Ovaltine labels," as I remember it). At the end of each episode of the *Captain Midnight* series on the radio, Pierre André, the announcer, would read off a series of numbers that *only we members* could understand. Important stuff like "Watch Shark."

As you may have guessed, our badges acted as decoders. A dial—much like the ones we find on combination locks—numbered one to twenty-six, sat inside a (fixed) circle that contained the letters of the scrambled alphabet. If the announcer said "tonight's code is k-4," we would rotate the dial accordingly, and *decrypt* the message by noting the letter opposite each number in the encrypted message.

For some reason, the Secret Squadron's nemeses—"the evil Ivan Shark and his daughter Fury"—never managed to capture one of our badges. Had they succeeded, Captain Midnight and his Ovaltine-drinking friends could no longer have communicated in secret. Each badge, after all, held the keys to the entire coding system.

Modern systems, on the other hand, offer better protection, as the following example illustrates.

Every CNN viewer recognized the key role that technology played in the 1991 Persian Gulf war. While missiles flew and planes carried out much photographed bombing runs, however, the war itself hinged upon a lightning-fast series of ground-based strikes that sent the Iraqi army fleeing northward in panic. In the end, armored cavalry wheeled through the desert in an old-fashioned flanking maneuver.

But here, too, technology helped decide the issue—albeit not in ways as apparent to television viewers. The item I have in mind got minimal press coverage. It wasn't photogenic, and the army wasn't all that anxious to publicize it. Still, a senior military officer attached to President Bush's staff cited it as the single most important weapon in allied hands.

What was it? Just a telephone. But a telephone with a special extra. When one officer used it to call another, a single button push at each end of the line ensured the absolute security of the ensuing conversation. As a result, an allied tank commander could shout his unit's position to supporting artillery in the heat of battle without the slightest concern about enemy eavesdroppers. Iraqi radios, on the other hand, provided a running gold mine of information to allied intelligence. Imagine the outcome of a battle in which only one side knows the other's location.

While anyone in possession of such a phone could place a private call to any other, the loss of one of these instruments to the enemy would have left all the others' security entirely intact. Why? In the case of the

Captain Midnight badges, the nested pair of letter and number circles allowed each possessor to *encrypt* (letters to numbers) and *decrypt* (numbers to letters) messages. The phones created encryption (speech into code) recipes—called *public key* systems—that hid the corresponding decryption (code into speech) recipe from its users. In practice, this allowed two phones at opposite ends of a link to exchange randomly selected encryption recipes while keeping the corresponding decryption recipes to themselves (and away from any eavesdroppers).

With the war over, and with the Soviet Union's collapse, suppliers of military technology naturally turned toward civilian markets. A secure telephone seemed a natural product for security-conscious people in many occupations, so several of the companies that had supplied secure phones to Desert Storm embarked on civilian versions.

But few such units reached ordinary consumers. Instead, the manufacturers received friendly suggestions that they switch to a high-performance encryption chip that the government had designed at its own expense. Government altruism? Not quite. While experts tell me that the new chip appears to make unauthorized decryption harder than most existing methods, the new chip comes with an electronic back door—a door to which only certain government agencies will hold the keys. Whether or not the U.S. government succeeds in this effort, however, it seems clear that the vast majority of tomorrow's citizens will just touch a button to shield their conversations from all but the world's most determined snoopers.

In addition, public-key encryption technology can offer consumers much more than merely protecting their conversation from eavesdroppers. Take electronic signatures. Suppose you wanted to send an electronic message to your stockbroker concerning a financial transaction. How would the broker know that the message came from you? Instead of shipping the encoding recipe to the broker, as you would if you wanted the broker to send *encrypted* messages back to you, you would ship the *decoder* instead. In this latter instance, you aim to prevent others from creating encrypted messages, but you don't care who reads them. Forgers, rather than eavesdroppers, concern us in this application.

As long as no one can deduce the encryption method from the corresponding (now public) decryption method, only the author can create messages that that particular decryptor can decipher. Suppose I wished to write checks on my bank via e-mail. I could send an encrypted file along with a clear-text message to PSE&G (my local electric company) saying, "Send this file to United Jersey Bank to collect the $83.87 that I owe you."

I would also send PSE&G a copy of the decryption method so they could verify the amount of my "check" before they deposited it in their account. But they would have to forward the encrypted version of my message in order to get paid. After all, anyone can write "Please pay $83.57 to PSE&G out of Arno Penzias's account." For that reason, my bank would also retain a copy of that message's encrypted version, just as they now preserve an image of my signed check, so as to prevent me from denying that I had "signed" it in the first place.

In addition to their general utility in moving away from paper, unforgeable electronic signatures may also offer us an interesting platform for privacy protection. Suppose I were to give a bank a certain amount of cash (or an electronic check, made anonymous by a scheme such as now follows) along with a decryption algorithm of the encryption/decryption pair I created for that account. In return, they would supply me with an account number and agree on a PIN code so I could check my balance over the phone. That's it. I need never give them my name or address. The encryption algorithm I devised would not only act as my signature, it would also connect me to the account without revealing my identity, acting as a *pseudonym* instead.

If I wanted a monthly statement, I'd merely supply the bank with an extra encryption key and ask that the encrypted statement be posted on one of the electronic bulletin boards that abound on the Internet. Gangsters and tax evaders need no longer worry about whether or not some Swiss bank might succumb to arm-twisting by the U.S. Internal Revenue. Small wonder that law enforcement agencies rue the advent of unbreakable ciphers.

While I could connect that *pseudonym* to myself by demonstrating my coding algorithm in case of a lawsuit against the bank, I could use it in place of my own name for whatever transactions I wished to keep private. When the bank in question got my "check" from PSE&G, it would have no way of finding out which of PSE&G's customers made the payment, short of cracking the code.

Whether or not such schemes catch on with the general public, criminals will hardly pass up such opportunities for concealing their activities. As a result, prudent business practice will surely demand greater attention to the prevention of information thefts—rather than trying to track them down afterward. Safeguarding information, therefore, may remain an optional consideration for everyday transactions, but it will surely take on increasing importance in the workplace.

While average citizens may well wish to shield at least some of their actions from public scrutiny—if for no other reason than to ward off overeager salespeople—each of us still remains responsible for his or her actions. Some banks may assign a certain financial worth to a faceless computer code, but that hardly amounts to a character reference. A good name still counts for something. As we bury our personae within encrypted envelopes, therefore, we should remain heedful of the difference between the rule-driven game playing at which machines excel and the socialized behavior that helps define our humanity. In that way, we can avoid the sterile consequences that can result from confusion between mind and machine.

Mechanomorphization?

Superficial similarities between computers and humans abound in today's world and will surely grow even more abundant in the future. As increasingly functional products and services play an ever bigger role in modern life, how can we make sure that human users won't come to emulate their helpers' behaviors rather than the other way around?

Look at computer-based education, for example. As one networked computer transfers programs and data to another, it imparts the knowledge represented in that software. In other words, a rudimentary form of "teaching" takes place between machines. At the same time, increasing numbers of people learn about the world by typing at keyboards and studying displayed outputs. How should producers of educational software distinguish between these two groups of students?

With much benefit—including hoped-for savings in labor costs— expected from machine-based teaching, computer-aided instruction to people seems a surefire growth industry. Two hundred years ago, American citizens formed opinions on the basis of then popular media, as exemplified by the *Federalist* papers, which were published in the newspapers of the day. Today, even modern graduate students rarely read these papers unless forced to. Fifteen-second sound bites have assumed the role of multipage documents. What has changed? Since biology can't move that fast, we can conclude that flexibly adaptive young brains have attuned themselves to the environment that present-day media presents to their sensory input.

As with other influences, the role of computers in each student's development deserves caring attention—and not just for young children. Even grown-ups who ought to know better sometimes let their infatuation with machines overwhelm their judgment.

In the course of writing this book, I encountered an eminent psychiatrist who predicted that many patients in need of counseling would someday converse with a computer program. "Best of all," he concluded, "it won't be troubled by countertransference." In other words, the machine won't get angry or aroused by anything the patient says or does.

Such electronic neatness makes me uneasy. Awareness of their own reaction to seemingly innocent phrases or gestures can give therapists—or marketing managers, for that matter—pointers to deeper meanings. From this perspective, replacing a human with a machine incapable of such responses diminishes the patient's—or customer's—ability to convey information. Rather than a cleaner pathway, therefore, this emotion-free system leads to a narrower channel. Whatever can't get through gets trashed. Information, I submit, is too valuable for us to strip away everything but the numbers.

From a business perspective, managers must guard against becoming too obsessed with the numbers—even though number-preoccupied businesses sometimes flourish in today's world. For example, Tom Peters and Robert Waterman's 1980s runaway best-seller, *In Search of Excellence,* extolled the virtues of a select group of companies, all chosen on the basis of a certain period of continuous growth in earnings. But a decade later, when Peters reviewed these narrowly based choices, he found marked splits in the pack's later performance—the effect of factors that numbers alone couldn't capture.

For some companies, a long string of financial successes attested to an ability to thrive in a stable and growing business but offered no clue as to their ability to enter new ones in response to changing circumstances—such as the switch from mainframes to networked computers.

In other cases, anecdotal evidence suggests too much of a break with the past rather than too little. With the focus on finances to the exclusion of everything else, these companies treated their component parts as nothing more than tradable engines for making money. Here, I believe, what organizational theorists describe as the "theory of the firm"—the sustainable competitive advantage an organization derives from the company-specific competence of its people—apparently dwindled away. Treated on the basis of numbers alone, employees began to act on that basis as well.

As managers look to computer-based information sources for help in making decisions, they risk facing the proverbial drunk's dilemma: Should he search for his lost coin on the dark roadway where he lost it or under a distant streetlight where he can see so much more clearly? Given computing's power to illuminate numerical problems, many managers seem tempted to conduct their search for answers in that arena alone, even though the real questions lie elsewhere.

Infatuation with early computers led their enthusiasts to dub these sterile number crunchers "electronic brains"—as if real brains had no other capabilities worth noting. How much more opportunity for myopia in the richer, more flexible technology outlined in these pages. As computers begin to speak and recognize faces, future generations of managers will need deep and abiding awareness of the difference between personhood and programming.

Machines, after all, can watch "ticker tapes" and hedge dollars against pounds or pesos when the spread allows room for a profitable transaction. But only human minds can grasp the real meaning behind those numbers—real economics, of real countries and real people. As computer-literate managers become increasingly adept with numbers, therefore, they must also nourish their understanding of the social and economic realities that bare data merely reflect.

Rich and Poor

With corporate globalization, systems integration, and the universal deployment of information all elevated to the rank of pieties by today's business and government leaders, few can doubt the likelihood of even closer business links among the world's advanced economies. But what about the rest? Will less developed countries fall even further behind as technology advances? In other words, will the harmony concept serve to divide the world, even as it unifies the economies that serve its most fortunate citizens?

Some hopeful signs exist. Within days after the January 1, 1994, start of their armed struggle against their government, southern Mexico's impoverished Chiapas Indians had posted their grievances and calls for support on electronic bulletin boards around the world. No need to wait for venturesome reporters to bring their side of the conflict to worldwide attention. Just get hold of a PC and call up an Internet access provider. No harder to obtain than a few extra rifles, and probably more effective in the long run. As this example suggests, at least some of the Harmony Era's benefits seem likely to make themselves felt across a wide income range.

From a technological perspective, there appears ample means for closing the gap between rich and poor. Since a growing fraction of the world's most valuable resources are based upon software, they don't suffer from physical limitations. So sharing these new, information-based sources of wealth can cost relatively little. Moreover "donors" may see gains in the form of expanded markets and unexpected applications of their products. The rapid movement of China and—more recently—India toward the mainstream of the world's market economies appears to lend credence to this conclusion.

As we've seen, access to technology increases the power of its processors vis-à-vis older forms of economic advantage. In particular, I feel safe in saying that most of the economies that expect to depend upon raw materials alone have little cause for optimism. While I wouldn't complain if someone gave me an oil well or a diamond mine, I see numerous opportunities for homegrown substitutes. Advanced economies such as those of France and Japan have reduced their dependence upon petroleum imports by adopting coherent approaches to the generation of electric power from nuclear energy, for instance, while countries like Great Britain and Norway have stretched their petroleum supplies by finding ways of squeezing extra oil from their deep ocean wells.

Similarly, the science section of my favorite newspaper informs me that laboratory techniques for producing synthetic diamonds keep advancing. And far more important reductions in our dependency upon raw material vendors—such as fabrication improvements that enable a 50 percent reduction in the wall thickness of aluminum beverage cans—happen so frequently that the general public rarely takes notice.

While the above examples touch upon physical commodities, much the same holds true for so-called low-priced labor. With careful attention to design of production processes, astute managers in developed countries can frequently overcome disparities in labor costs. For instance, one might imagine that the advent of worldwide data networking would have moved large numbers of America's computer-programming jobs to countries like India, where well-educated English-speaking programmers earn only a fraction of U.S. salaries. But no such move has taken place. While several agencies offer such service to smaller users—and some larger companies have experimented with branch operations—offshore programming has yet to make more than a tiny dent in that industry. Why?

Software manufacture consists of a sequence of processes, of which the actual writing of the code itself represents less than one-third of the total task. Most of the job consists of finding out what the customer wants and making sure that the program performs as required. And both these functions generally require interaction with customers to resolve ambiguities. By the time a software provider has described the remaining portion with sufficient precision for an overseas handoff, little value remains to be added. In some cases, a computer can even generate the desired program from specifications without the aid of manual intervention. Unlike the manufacture of more easily specified items, such as clothing or dinnerware, software production shows little sign of moving to low-wage countries.

On the other hand, integrated-circuit (IC) design generally takes place along more geographically distributed lines. Most IC manufacturers cluster their production facilities in as few locations as possible to justify the massive capital expenditures associated with each site. Designers, on the other hand, can work in relatively small teams on individual chip designs. So chip manufacturers tend to distribute design centers close to present and potential customers—somewhat like information-age vending machines for high-function custom electronic circuitry.

In this environment, places like Israel appear to provide attractive locations for IC design centers, at least judging by the impressive list of U.S. chip makers who maintain sizable operations in that country. While much of Israel's chip design finds its way into products made on other continents, a significant fraction goes to Israeli factories—not only increasing the local labor content of the finished product but also giving local equipment manufacturers better access to the specialized chips they need. A nice boost for that country's economy, I'd say.

But not everyone agrees. "Potato chips or silicon chips. What's the difference? Chips are chips," a skeptical member of the Bush administration is said to have remarked when pressed about his reluctance to give special attention to maintaining U.S. presence in high-tech manufacture. From this official's perspective, the dollar value of a nation's output matters far more than the mix of products and services that produced that output.

Clearly, an influential school of thought ties economic progress to investment, while (as even a casual reader of these pages should realize by now) I tie it to innovation. Since neither school holds a monopoly on fundamental truth, I leave it to interested readers to get the other side of the story elsewhere. From my perspective, I see no reason to expect any near-term rupture in the traditional link between manufacture and economic prosperity that has held sway since the dawn of the Industrial Revolution. Certainly, the economic role of intellectual property will continue to increase in importance, but the economic return on intellectual property generally comes through the product made—whether a jet airliner or a pair of basketball shoes. Just think how much more a half ton of steel is worth when offered in the form of a Mercedes than a Trebant (the former East Germany's much ridiculed jalopy), for instance.

I see enormous growth potential for integrated circuits in the future. Indeed, all but the most elaborate of our present-day electronic artifacts will surely shrink to single-chip size, or even smaller. Look at where we stand in semiconductor memories. The first DRAM (dynamic random access memory) chips could hold four thousand data bits. At the time of this writing, the four megabit chip acts as the storage workhorse of our electronics industry. And progress continues. Based upon just the expected evolution of present-day design and fabrication technology, my integrated-circuit research colleagues assure me that four gigabit chips will appear in the marketplace in the early years of the next decade.

Think of it. At four million bits per chip, today's DRAMs provide one thousand times the storage capacity of the four kilobit chip that started the sequence. With four *billion* bits per chip in prospect, we can look forward to yet another factor of one thousand.

What does this mean in practical terms? Picture the largest consumer electronics store you've ever seen or heard about, with row upon row of TV sets, stereos, personal computers, electronic games, VCRs, camcorders, and the like. In the four gigabit era, all the electronic circuits in every appliance in sight will fit into barely enough silicon to make a noticeable bulge in a shirt pocket.

Consider how much design and assembly effort each such chip will represent a few years from now. While unit prices of individual functions will surely plummet as volume grows, these crucial piece parts seem destined to constitute an increasing share of manufacturing value. Most of the so-called *assembly* factories I know about today use less and less space to produce more and more output, because most of the assembly work they once carried out now comes prepackaged in the purchased components they put together.

Back in the 1960s, Fidel Castro was said to have banned the importation of aluminum cookware so as to spur local manufacture. "Nice try, but no cigar," as they say. Aluminum ingots actually cost more per pound than pots and pans do—at least in the volumes that the Cubans wished to buy. Although the notion failed in this instance, prudent governments can't afford to overlook the employment opportunities that domestic manufacturing generally offers.

Unfortunately for Dr. Castro's regime, Cuban manufacturers couldn't overcome the advantages of scale—notably advantageous access to raw materials—held by their far larger competitors. But at least some of today's hopefuls may stand a better chance. In the software-coding case I cited earlier, would-be programmers hope to insert themselves in the middle of a multistage process, one in which most of the value gets added before the actual coding even begins.

By contrast I see more promising opportunities for "ground-floor" entry into value creation through local IC design, thanks to the growing dispersal of computer-aided design around the globe. Since the factories that produce ICs require massive investments, earnings-conscious managers must strive to keep their production lines filled. By serving a diverse customer base, they can cushion themselves against fluctuations. Look at IBM's emergence as an electronic component vendor. Still the world's largest IC producer, this manufacturing giant now sees advantage in no longer restricting component sales to its internal operations.

As electronic circuits become cheaper, moreover, they will likely subsume an increasing portion of the role now played by mechanical technology. Remember when mechanical timers ran most home appliances? Just as the plummeting cost of digital clocks changed that picture, note that an electret microphone now costs less than a set of push buttons. As cheaper and more powerful circuitry becomes commonplace, speech recognition should cost no more than an electronic timing circuit. In many applications, talking may not only be more convenient than typing, it may also be cheaper.

Under such circumstances, appliance manufacturers can ill afford to thumb through component catalogs to realize such economies. They must provide for such functionality in their components and incorporate them into an overall design at the outset. Otherwise, they leave themselves with little more than a sales outlet role for someone else's products.

Fortunately, the microprocessor's overthrow of centralized computing promises to bring supercomputer-sized data processing to the desktop—wherever that desktop might be. As a result, electrical and mechanical design capabilities should move easily across geographic barriers. And geographically dispersed design teams can take advantage of the growing communications capabilities of their workstations as well. That way, a group in one country can design a sophisticated subassembly in one place and assure its precise fit with one designed in another place.

While even the most optimistic of future scenarios cannot overlook the enormous advantages today's developed economies enjoy vis-à-vis their less fortunate neighbors, the Harmony Era that looms before us should offer significant opportunities for economic change. We have seen how yesterday's mechanically driven production lines favored massive numbers of the same product. Many of today's factories, on the other hand, cheerfully accept varied products. So cheerfully, in fact, that some manufacturing engineers contemplate so-called lot-size-of-one scenarios—factories that retool themselves on a per-item basis.

At the other end of the product-delivery chain, I fully expect tomorrow's integrated systems—like the one portrayed in Michael's "home-helper" scenario—to accommodate products from multiple providers. While these paths to entry don't guarantee success, they at least allow new players to join in the game.

No single producer can hope to provide the diverse functionality that tomorrow's users will demand. So tomorrow's systems will surely provide lots of opportunities for add-ons—ranging from special-purpose robots to electronic games—from multiple vendors. These vendors, in turn, should be able to locate their operations anywhere they can find properly trained people. Computer-based design tools together with desktop communication should allow such vendors to forge effective links with suppliers, partners, and potential customers. Geographic separation should therefore offer few obstacles to entrepreneurs who seek to participate in this emerging market for sophisticated artifacts.

Hopefully, this global diversity of offerings will result in new products and services that will grow new markets. As a result, users can enjoy enhanced lifestyles, while newly enabled entrepreneurs can gain profits for themselves and employment for their fellow citizens. Making it happen, of course, remains up to all of us. Citizens of economically advantaged countries have much to gain from continued growth in world trade and the consequent bridging of the gaps that now separate rich from poor.

Harmony, then, aims at a continuing reduction in discord. In the technology arena, this concept addresses the gaps that separate human handiwork from its users, one component from another, and much of that handiwork from nature itself. Viewed from a business perspective, this translates into new opportunities for commerce and employment. But we can go further. A truly healthy economy—healthy for everyone, not just for a fortunate few—demands that we employ people and technology to bridge gaps within human society itself.

In that sense, the picture of a network carrying sophisticated computer designs across oceans and continents—while also serving as a global forum for the grievances of impoverished villagers—provides a graphic example of the use of human ingenuity for human benefit. Harmony, after all, is what we make it.

INDEX

people-with-machines, 100–101, 105–7
people-with-people, 100, 101–5
Dirksen, Everett, 54
Downsizing, 5, 8, 14, 49, 52, 69
DRAM (dynamic random access memory), 166
Drucker, Peter, 62
Duquesne Light, 141

Eastern Europe, 133, 134
Economies of scale, 22–23, 24, 25, 26, 35, 73
Education, 56
  machine-based teaching, 161
Einstein, Albert, 57
Einstein's theory of relativity, 19
Eisenhower, Dwight D., 141
Electronic mail (E-mail), 79–81, 89, 92, 154
Electronic signature, 159–60
Employment
  criteria in selecting an employer, 5–6
  downsizing, 5, 8, 14, 49, 52, 69
  full employment, xiii
  future technological advances and, 49–50, 75–77
  revenue increases necessary to maintain current levels of, 48
Encryption, 157–60
Environmental concerns, 64, 131–47
  in Eastern Europe, 133, 134
  economic concerns and, xiii, xiv
  fish-tagging technology, 132–33
  in the future, 146–47
  green manufacturing, 134–37
  nuclear power, 139–45
  recycling, 137–39, 142
  technology at odds with, 12
  transportation and, 145–46
Environmental Protection Agency (EPA), 135
Esthetics, 65
Europe, 56
  environmental concerns treated as opportunities in, 131

unemployment in, 9
unification of, 8–9

Fax machines, 112
Federal Aviation Administration (FAA), 97–98
Film studio, 63
Fish-tagging technology, 132–33
Fission, 140
Ford, Henry, 37–38
FORTRAN, 21
Fortune, 67
Fortune 500 companies, 21
  downsizing by, *see* Downsizing
Foster, Robert, 6, 7
France, 143–44, 164
Fuel cells, 144–45
Fujitsu SX-3, 20
Full employment, xiii
Fundamental science, 19
Fusion, 140
Fusion Science Institute, Japan, 20

Galbraith, John Kenneth, 22, 23
Gaps in information, businesses filling, 7–8, 58–60
Gates, Bill, 121–22
General Accounting Office (GAO), 145
General Electric (GE), 141
General Foods, 44
General Motors, 22, 26, 73
General Sciences Administration, 97–98
Germany, 143
Gesture-based inputs, 100–101
Ginott, Chaim, 116
Glass fibers, 89–91, 92
Global economy, 36
Gore, Al, 81
Great Britain, 134, 164
Green manufacturing, 134–37

Harlem, New York, 85
Harmony, 1, 13

Usability (*con't.*)
   people-with-machine communications, 100–101, 105–7
   people-with-people communications, 100, 101–5

Value-added resellers (VARs), 110
VCRs, 105
Vendor-customer relationships, 45
Video communications, 101–2
Videoconferencing, 56
Virtual-circuit concept, 94–95
Voice/data/video merger, 94–96
Volkswagen, 139
Von Neumann, John, 18

Wal★Mart, 40–42, 88
Walton, Sam, 40
Wang, 52
War and Remembrance (Wouk), 68–69

Watches, 9–10
Waterman, Robert, 162
Weapons
   design of, 19–20, 55
   see also Nuclear power
Weather forecasting, 20
Westinghouse, 52, 141
Whyte, William H., 34
Wireless technology, 91
Workstations, 39
Workstyles
   emerging, 71–77
   looking sideways, 71, 72
   see also Organizational structure
Worldwatch Institute, 143, 144, 146
Wouk, Herman, 68–69
Wozniak, Steve, 27
Wriston, Walter, 36, 43

Xerox, 27, 58